CLIMATE CHANGE IMPACTS ON THE STABILITY OF SMALL TIDAL INLETS

CLIMATE CHANGE IMPACTS ON THE STABILITY OF SMALL TIDAL INLETS

DISSERTATION

Submitted in fulfilment of the requirements of

the Board for Doctorates of Delft University of Technology

and

of the Academic Board of the UNESCO-IHE Institute for Water Education

for

the Degree of DOCTOR

to be defended in public

on Tuesday, 1st December 2015, at 15:00 hours

in Delft, the Netherlands

by

Trang Minh Duong

Master of Science, UNESCO-IHE, Delft, the Netherlands
born in Hanoi, VietNam

This dissertation has been approved by the promoters:
Prof. dr. ir. J. A. Roelvink and Prof.dr. R.W.M.R.J.B. Ranasinghe

Composition of the Doctoral Committee:

Chairman	Rector Magnificus, TU Delft
Vice-Chairman	Rector, UNESCO-IHE
Prof. dr. ir. J.A. Roelvink	UNESCO-IHE/Delft University of Technology, Promotor
Prof. dr. R.W.M.R.J.B. Ranasinghe	UNESCO-IHE/Australian National University, Promotor

Independent members:
Dr. S. Weesakul	Asian Institute of Technology/HAII, Bangkok, Thailand
Prof. dr. M. Larson	Lund University, Sweden
Prof. dr.ir. A.E. Mynett	UNESCO-IHE/Delft University of Technology
Prof. dr.ir. Z.B. Wang	Delft University of Technology
Prof. dr. D.P. Solamatine	Delft University of Technology, reserve member

CRC Press/Balkema is an imprint of the Taylor & Francis Group, an informa business

Published by:
CRC Press/Balkema
PO Box 11320, 2301 EH Leiden, The Netherlands
e-mail: Pub.NL@taylorandfrancis.com
www.crcpress.com – www.taylorandfrancis.com

ISBN 978-1-138-02944-6 (Taylor & Francis Group)

This research was supported by the DUPC-UPARF cooperation program between the Dutch Foreign Ministry (DGIS) and UNESCO-IHE and the AXA Research Fund.

ABSTRACT

The coastal zones in the vicinity of tidal inlets, which are commonly utilized for navigation, sand mining, waterfront developments, fishing and recreation, are under particularly high population pressure. The intensive population concentration and excessive natural resources exploitation in these areas could lead to biodiversity loss, destruction of habitats, pollution, as well as conflicts between potential uses, and space congestion problems, which will only be exacerbated by foreshadowed climate change (CC). Although a very few recent studies have investigated CC impacts on very large tidal inlet/basin systems, the nature and magnitude of CC impacts on the more commonly found small tidal inlet/estuary systems remains practically un-investigated to date. These relatively small estuaries/lagoons (also known as "bar-built" or "barrier" estuaries, and hereon referred to as Small Tidal Inlets or STIs) are common along wave-dominated, microtidal mainland coasts comprising about 50% of the world's coastline.

Due to their common occurrence in the tropical and sub-tropical zones, most STIs are found in developing countries, where data availability is generally poor (i.e. data poor environments) and community resilience to coastal change is low. Furthermore, STI environs in developing countries especially host a number of economic activities (and thousands of associated livelihoods) which contribute significantly to the national GDPs. The combination of pre-dominant occurrence in developing countries, socio-economic relevance and low community resilience, general lack of data, and high sensitivity to seasonal forcing makes STIs potentially very vulnerable to CC impacts and thus a high priority area of research. This study was therefore undertaken with the overarching objective of (a) developing methods and tools that can provide insights on potential CC impacts on STIs, and (b) demonstrating their application to assess CC impacts on the main types of STIs.

Throughout this Thesis, 3 case study STIs representing the 3 main STI Types are used:
- Negombo lagoon, Sri Lanka: Permanently open, locationally stable inlet (Type 1)
- Kalutara lagoon, Sri Lanka: Permanently open, alongshore migrating inlet (Type 2)
- Maha Oya river, Sri Lanka: Seasonally/Intermittently open, locationally stable inlet (Type 3)

To circumnavigate the inability of contemporary process based coastal area morphodynamic models to accurately simulate the morphological evolution of STIs over typical CC impact assessment time scales (e.g. 100 yrs) with concurrent tide, wave and riverflow forcing, 2 different snap-shot modelling approaches for data poor and data rich environments are proposed. The data poor approach uses schematized flat bed bathymetries that follow real world STIs and CC forcing derived from freely available coarse resolution global models while the data rich approach requires detailed bathymetries and downscaled CC forcing. Furthermore, to enable rapid assessments of CC impacts on STI stability, particularly to aid frontline coastal zone managers/planners, a reduced complexity model is developed based on existing knowledge and physical formulations. The model, which is capable of simulating 100 years in under 3 seconds on a standard PC, provides predictions of STI stability based on the Bruun inlet stability criterion r (= P/M; where P = tidal prism (m^3) and M = annual longshore transport (m^3/yr)).

Based on process based snap-shot model applications under contemporary forcing, a clear link between STI Type and r is established (Table 1).

Table 1. Classification scheme for inlet Type and stability conditions.

Inlet Type	r =P/M	Bruun Classification
Type 1	> 150	Good
	100 - 150	Fair
	50 - 100	Fair to Poor
	20 - 50	Poor
Type 2	10 - 20	Unstable (open and migrating)
Type 2/3	5 - 10	Unstable (migrating or intermittently closing)
Type 3	0 - 5	Unstable (intermittently closing)

All 3 modelling approaches show that Type 1 and Type 3 STIs will not change Type by the year 2100. For Type 2 STIs, the data poor approach suggests a Type change to Type 1 when CC results in a decreased annual longshore sediment transport, while the other two approaches predict no Type change under any CC forcing scenario. However, the results of the data rich approach and the reduced complexity model are likely to be more reliable due to the use of more accurate bathymetric data and site specific, downscaled CC forcings therein.

Although CC driven STI Type changes appear to be rather unlikely in the 21st century, model results do show that CC is likely to change the level of stability of STIs, indicated by significant future changes of the r value from its present value. At Type 1 STIs, future CC driven increases/decreases in longshore sediment transport may result in decreases/increases in their level of stability. At Type 2 and Type 3 STIs concurrent increases (decreases) in longshore sediment transport and decreases (increases) in riverflow may result in decreasing (increasing) the level of inlet stability. Sea level rise (SLR) appears not to be the main driver of change in the level of STI stability, with CC driven variations in wave direction emerging as the major driver of potential change in STI stability.

For future CC impacts assessment at STIs, an initial assessment using the reduced complexity model is recommended. If Type changes are predicted at any time (or if r drops below 10 for a Type 2 STI), or if specific insights (e.g. migration distance at Type 2 STIs, inlet closure time at Type 3 STIs) are desired, then it is essential that the (data poor or data rich, depending on which is feasible in the study area) process based snap-shot modelling approach be adopted.

SAMENVATTING

Kustgebieden in de nabijheid van zeegaten, die vaak worden gebruikt voor scheepvaart, zandwinning, waterfront ontwikkeling, visserij en recreatie, kennen een bijzonder hoge bevolkingsdruk. De hoge bevolkingsdichtheid en het verregaande gebruik van het natuurlijke system in deze gebieden kan een verlies van biodiversiteit, vernietiging van habitats en vervuiling tot gevolg hebben alsmede leiden tot conflicten in gebruik en overbelasting van de ruimte. Deze aspecten worden bovendien versterkt door de voorspelde klimaatverandering (KV). Hoewel een aantal recente onderzoeken de invloed van KV op grote zeegat/bekken systemen hebben onderzocht, is de aard en de omvang van de KV op kleine zeegat/bekken systemen tot op heden nauwelijks onderzocht. Deze relatief kleine estuaria/lagunes (ook wel 'bar built' of 'barrier' estuaria genoemd, hierna aangeduid als Kleine Zeegat Systemen of KZS) zijn wijdverspreid langs golf-gedomineerde, micro-getijde kusten langs de continenten, die grofweg 50 % van 's werelds kustlijn beslaan.

Doordat deze KZS veel voorkomen in tropische en subtropische gebieden, zijn de meeste KZS gelegen in ontwikkelingslanden waar de beschikbaarheid van gegevens over het algemeen laag is (zgn. data schaarse omgevingen) en er weinig veerkracht van de lokale gemeenschap is voor kustverandering. Bovendien huisvesten de gebieden rondom KZS, met name in ontwikkelingslanden, een aantal economische activiteiten die aanzienlijk bijdragen aan het bruto binnenlands product. De combinatie van dit voornamelijk voorkomen in ontwikkelingslanden, het sociaaleconomische belang, de beperkte veerkracht van de lokale bevolking, het gebrek aan gegevens en de gevoeligheid voor seizoensvariatie in de aandrijvende krachten zorgt ervoor dat de KZS in potentie zeer kwetsbaar zijn voor KV wat dit een onderzoeksgebied maakt met een hoge prioriteit. Dit onderzoek heeft derhalve de volgende overkoepelende onderzoeksdoelen: a) het ontwikkelen van methoden en instrumenten die inzicht kunnen geven in de mogelijke impact van KV op KZS, en b) het aantonen van de bruikbaarheid van deze methoden en instrumenten om de impact van KV op de belangrijkste typen KZS te beoordelen.

In deze dissertatie zijn 3 casestudy's van KZS onderzocht, die kenmerkend zijn voor de 3 belangrijke typen KZS:
- Negombo lagoon, Sri Lanka: Permanent open, plaatsvast zeegat (Type 1)

- Kalutara lagoon, Sri Lanka: Permanent open, kustlangs verplaatsend zeegat (Type 2)
- Maha Oya river, Sri Lanka: Seizoensgebonden/afwisselend open, plaatsvast zeegat (Type 3)

Om de beperkingen van de huidige proces gebaseerde kustmorfodynamica modellen om accuraat de morfologische verandering van KZS over de tijdschaal van KV (bv. 100 jaar) met de samenhangende aandrijvende krachten van getijde, golven en de rivier te overkomen, zijn twee verschillende snapshot model benaderingen voorgesteld voor zowel data schaarse en data rijke gebieden. De data schaarse benadering maakt gebruik van geschematiseerde vlakke bodemliggingen die werkelijke KZS volgen en KV aandrijvende krachten die zijn afgeleid uit vrij verkrijgbare, wereld dekkende modellen met een grove resolutie. De data rijke benadering vereist daarentegen gedetailleerde bodemligging en een beschrijving van de aandrijvende krachten van KV op kleinere schaal. Daarnaast is een model benadering met gereduceerde complexiteit ontwikkeld, gebaseerd op bestaande kennis en fysieke relaties, om een snelle beoordeling van KV impact op KZS mogelijk te maken, in het bijzonder om praktijkgerichte kustzone managers/planners te helpen. Het model, waarmee het mogelijk is om 100 jaar vooruit te voorspellen met een rekentijd van 3 seconden op een standaard PC, geeft voorspellingen van de stabiliteit van KZS door gebruik te maken van het Bruun zeegat stabiliteits criterium r (= P/M; waarbij P = getijde prisma (m^3) en M = jaarlijks kustlangs transport (m^3/jaar) is).

Door gebruik te maken van de proces gebaseerde snapshot model toepassingen met hedendaagse forcering is een duidelijke link vastgesteld tussen het type KZS en parameter r (Tabel 1).

Tabel 1. Classificatie schema voor het type zeegat en de stabiliteitscondities.

Type zeegat	r =P/M	Bruun Classificatie
Type 1	> 150	Goed
	100 - 150	Redelijk
	50 - 100	Redelijk tot Matig
	20 - 50	Matig
Type 2	10 - 20	Instabiel (open en kustlangs verplaatsend)
Type 2/3	5 - 10	Instabiel (kustlangs verplaatsend of afwisselend sluitend)
Type 3	0 - 5	Instabiel (afwisselend sluitend)

Alle drie de model benaderingen laten zien dat KZS van Type 1 en Type 3 niet van type zullen veranderen voor het jaar 2100. Voor KZS van Type 2 geeft de data schaarse aanpak aan dat het type van het KZS verandert naar Type 1 als KV leidt tot een afname van het jaarlijks kustlangse sediment transport, terwijl de overige twee modelbenaderingen geen type verandering voorspellen onder elk van de KV scenario's. Hierbij moet worden aangemerkt dat de resultaten van het data rijke model en het gereduceerde complexiteit model waarschijnlijk betrouwbaarder zijn omdat deze gebruik maken van nauwkeurigere gegevens van de bodemligging en gebied specifieke, KV gerelateerde aandrijvende krachten op kleinere schaal.

Hoewel het onwaarschijnlijk lijkt dat KV leid tot verandering van het type van KZS in de 21st eeuw, laten de model resultaten zien dat KV wel de mate van stabiliteit van KZS verandert, wat zichtbaar is als een significante verandering in de r waarde ten opzichte van de huidige waarde. Voor Type 1 KZS kan een KV gedreven toename/afname van het kustlangse sediment transport leiden tot een afname/toename van het stabiliteitsniveau. Voor KZS van Type 2 en Type 3 kunnen gelijktijdige toename (afname) in het kustlangs sediment transport en afname (toename) in de rivierafvoer leiden tot een afnemende (toenemende) stabiliteit van het zeegat. Zeespiegelstijging lijkt niet de belangrijkste drijvende kracht voor het stabiliteitsniveau van de het KZS, maar KV gedreven veranderingen in de golfrichting kwamen naar voren als de belangrijkste drijvende kracht voor potentiele verandering in de stabiliteit van KZS.

Voor een toekomstige beoordeling van de impact van KV bij KZS, wordt aangeraden een eerste beoordeling te maken met het model met gereduceerde complexiteit. Indien verandering van het zeegat type wordt voorspeld gedurende de looptijd van de voorspelling (of indien r waardes onder de 10 komen voor KZS van Type 2), of indien specifieke inzichten benodigd zijn (bv. de kustlangse verplaatsingsafstand bij KZS van Type 2, of de tijd tot afsluiting van het zeegat bij KZS van Type 3,), dan is het essentieel om de (data schaarse of data rijke, afhankelijk van wat haalbaar is in het interesse gebied) proces gebaseerde snapshot model benadering toe te passen.

This abstract is translated from English to Dutch by Dr.ir. Matthieu de Schipper, Faculty of Civil Engineering and Geosciences, Delft University of Technology.

Contents

CHAPTER 1

INTRODUCTION

1.1   Problem statement

A tidal inlet is defined as a waterway connection between the ocean and a protected embayment, it may be a bay, lagoon, or estuary through which tidal and other currents flow (Carter, 1988). Thus, it is through the inlet that the exchange of water, sediment and pollutants occur between the ocean and the lagoon/estuary.

Tidal inlets are of great societal importance as they are often associated with ports and harbours, industry, tourism, recreation and prime waterfront real estate. Tidal inlets are also among the most morphologically dynamic regions in the coastal zone (Kjerfve, 1994; Nicholls et al., 2007; Ranasinghe et al., 2013). The complex feedbacks between system forcing and response in these areas result in ongoing spatial and temporal variations in system characteristics which are of great scientific interest and continue to be the focus of numerous scientific studies (O'Brien, 1931; Escoffier, 1940; Bruun, 1978; Aubrey and Weishar, 1988; Prandle, 1992; Ranasinghe and Pattiaratchi, 2003; FitzGerald et al., 2008; Bertin et al., 2009; Lam, 2009; van der Wegen et al., 2010; Bruneau et al., 2011; Tung, 2011; Nahon et al., 2012; Dissanayake et al., 2012; Ranasinghe et al., 2013).

Tidal inlet behaviour is governed by the delicate balance of oceanic processes such as tides, waves and mean sea level (MSL), and fluvial processes such as riverflow and fluvial sediment fluxes. Alarmingly, all of these processes can be affected by climate change (CC) processes, which may result in severe negative physical impacts such as erosion of open coast beaches adjacent to the inlet and/or estuary margin shorelines, permanent or frequent inundation of low lying areas on estuary margins, eutrophication, and toxic algal blooms etc. Furthermore, CC driven changes in forcing may affect the stability of the inlet itself. For example, a permanently open, locationally stable inlet may evolve into an alongshore migrating, intermittently closing inlet or a seasonally closing, locationally stable inlet may evolve into a permanently open, alongshore migrating inlet. Such changes in inlet condition are highly likely to affect navigability and estuary/lagoon water quality leading to significant socio-economic, environmental and ecological losses.

Although a very few recent studies have investigated CC impacts on very large tidal inlet/basin systems (e.g. Dissanayake et al., 2012; van der Wegen, 2013), the exact nature and magnitude of CC impacts on the more commonly found small tidal inlet/estuary systems remains practically un-investigated to date. These relatively small estuaries/lagoons, which are also known as "bar-built" or "barrier" estuaries (hereafter referred to as Small Tidal Inlets or STIs for convenience) are common along wave-dominated, microtidal mainland coasts comprising about 50% of the world's coastline (Ranasinghe et al., 2013). While the exact number of STIs present around the world is unknown, it is likely to run into thousands with predominant occurrence in tropical and sub-tropical regions (e.g. India, Sri Lanka,Vietnam, Florida (USA), South America (Brazil), West and South Africa, and SW/SE Australia). STIs generally have little or no intertidal flats, backwater marshes or ebb tidal deltas. The barrier of these systems is usually a sand spit that is connected to the mainland, in contrast to a barrier island where the barrier is completely separated from the mainland. STI systems usually contain inlet channels that are less than 500m wide connected to relatively shallow (average depth < 10 m) estuaries/lagoons with surface areas less than 50 km^2.

There are 3 main STI Types:
- Permanently open, locationally stable inlets (Type 1)
- Permanently open, alongshore migrating inlets (Type 2)
- Seasonally/Intermittently open, locationally stable inlets (Type 3)

The most severe CC impact on a given STI would be a change in Type. This could potentially affect all or most economic and social activities centred on the STI that had developed over time based on the expectation that the general morphological behaviour of the STI will remain unchanged. For example, if a Type 1 system of which the lagoon is used as an anchorage for sea going vessels changes into a Type 3 system, it may no longer be possible to operate as an efficient anchorage. A less severe, but still potentially very socio-economically damaging CC impact would be a significant change in the level of stability of an STI (as per the Bruun inlet stability criterion $r = P/M$; where P = tidal prism (m^3) and M = annual longshore transport (m^3/yr); Bruun, 1978), while its Type remains unchanged. For example, if the level of stability of the same example Type 1 STI changes from 'good' ($r > 150$) to 'poor' ($20 < r < 50$), although it will still remain as a

Type 1 inlet, navigation through the inlet might become difficult and perilous, thus seriously compromising its continued functionality as an efficient anchorage.

Due to their pre-dominant occurrence in tropical and sub-tropical zones, STIs usually experience a strong seasonal cycle in system forcing comprising high and low riverflow/wave energy seasons (monsoon/non-monsoon or winter/summer). Their common occurrence in the tropical and sub-tropical zones also results in most STIs being found in developing countries, where data availability is mostly poor (i.e. data poor environments) and community resilience to coastal change is rather low compared to that in developed countries. Furthermore, STI environs in developing countries especially host a number of economic activities (and thousands of associated livelihoods) such as harbouring sea going fishing vessels, inland fisheries (e.g. prawn farming), tourist hotels and tourism associated recreational activities which contribute significantly to the national GDPs. The combination of pre-dominant occurrence in developing countries, socio-economic relevance and low community resilience, general lack of data, and high sensitivity to seasonal forcing makes STIs potentially very vulnerable to CC impacts and thus a high priority area of research. This Thesis therefore entirely focusses on CC impacts on STIs.

1.2   Objective and Research questions

1.2.1   Objective

The overarching objective of this study is to (a) develop methods and tools that can provide insights on potential CC impacts on STIs, and (b) to demonstrate their application to qualitatively and quantitatively assess CC impacts on different types of STIs.

1.2.2   Research Questions

To achieve the above objective, this study will attempt to answer the following specific research questions:

- *Research Question 1:* Can a process based coastal area model be used to assess CC impacts on STIs?
- *Research Question 2:* Can an easy-to-use reduced complexity model be developed to obtain rapid assessments of the temporal evolution of STI stability under CC forcing?
- *Research Question 3:* Is there a link between STI Type and the Bruun inlet stability criterion ($r = P/M$; where P = tidal prism (m^3) and M = annual longshore transport (m^3/yr)) that could aid in classifying different STI responses to CC?

- *Research Question 4:* Will CC change STI Type?

- *Research Question 5:* How will CC affect the level of stability of STIs?

- *Research Question 6:* What basic guidelines can be given to coastal zone managers on how to assess CC impacts on STIs to inform CC adaptation strategies?

References

Aubrey, D. G. and Weishar, L., (Eds), 1988. Hydrodynamics and sediment dynamics of tidal inlets. Lecture Notes on Coastal and Estuarine Studies, 29. Springer-Verlag, New York. 456p.

Bertin, X., Fortunato, A.B., Oliveira, A., 2009. A modeling-based analysis of processes driving wave-dominated inlets. Continental Shelf Research, 29 (5–6), 819–834.

Bruneau, N., Fortunato, A.B., Dodet, G., Freire, P., Oliveira, A., Bertin, X., 2011. Future evolution of a tidal inlet due to changes in wave climate, sea level and lagoon morphology: O'bidos lagoon, Portugal. Continental Shelf Research 31, 1915–1930.

Bruun, P., 1978. Stability of tidal inlets – theory and engineering. Developments in Geotechnical Engineering. Elsevier Scientific, Amsterdam, 510p.

Carter, R. W. G., 1988. Coastal Environments. Academic Press. London. 617p.

Dissanayake P. K.., Ranasinghe, R., Roelvink, D., 2012. The morphological response of large tidal inlet/basin systems to sea level rise. Climatic Change, 113, 253-276.

Escoffier, F.F., 1940. The stability of tidal inlets. Shore and Beach, 8,111–114.

FitzGerald, D.M., Fenster, M.S., Argow, B.A., Buynevich, I.V., 2008. Coastal impacts due to sea-level rise. Annual Review of Earth and Planetary Sciences, 36, 601-647.

Kjerfve, B., 1994. Coastal Lagoon Processes. In: Kjerfve, B., (Ed), Coastal Lagoon Processes. Elsevier Science Publishers, Amsterdam, pp. 1-8.

Lam, N. T., 2009. Hydrodynamics and morphodynamics of a seasonally forced tidal inlet system. Ph.D. Thesis, Delft University of Technology.

Nahon, A., Bertin, X., Fortunato, A.B., Oliveira, A., 2012. Process-based 2DH morphodynamic modeling of tidal inlets: A comparison with empirical classifications and theories. Marine Geology, 291–294, 1–11, doi:10.1016/j.margeo.2011.10.001.

Nicholls, R.J., Wong, P.P., Burkett, V.R., Codignotto, J.O., Hay, J.E., McLean, R.F., Ragoonaden, S., Woodroffe, C.D., 2007. Coastal systems and low-lying areas. Climate Change 2007: Impacts, Adaptation and Vulnerability, Contribution of Working Group II to the Fourth Assessment Report of the Intergovernmental Panel on Climate Change, Cambridge University Press, Cambridge, UK.

O'Brien, M.P., 1931. Estuary and tidal prisms related to entrance areas. Civil Engineering 1(8), 738-739.

Prandle, D., 1992. Dynamics and Exchanges in Estuaries and the Coastal Zone. American Geophysical Union, Washington. 650p.

Ranasinghe, R., Pattiaratchi, C., 2003. The seasonal closure of tidal inlets: causes and effects. Coastal Engineering Journal, 45(4), 601-627.

Ranasinghe, R., Duong, T.M., Uhlenbrook, S., Roelvink, D., Stive, M., 2013. Climate change impact assessment for inlet-interrupted coastlines. Nature Climate Change, 3, 83-87, DOI.10.1038/NCLIMATE1664.

Tung, T. T., 2011. Morphodynamics of Seasonally closed coastal inlets at the central coast of Vietnam. Ph.D. Thesis, Delft University of Technology.

van der Wegen, M., Dastgheib, A., Roelvink, J.A., 2010. Morphodynamic modeling of tidal channel evolution in comparison to empirical PA relationship. Coastal Engineering. 57, 827–837, doi:10.1016/j.coastaleng.2010.04.003.

van der Wegen, M., 2013. Numerical modeling of the impact of sea level rise on tidal basin morphodynamics, Journal of Geophysical Research, 118, doi:10.1002/jgrf.20034.

CHAPTER 2

ASSESSING CLIMATE CHANGE IMPACTS ON THE STABILITY OF SMALL TIDAL INLET SYSTEMS: WHY AND HOW?

2.1   Introduction

Coastal zones have historically attracted humans and human activities due to their amenity, aesthetic value and diverse ecosystems services, resulting in rapid expansions in settlements, urbanization, infrastructure, economic activities and tourism. At present, approximately 23% of the global population lives within 100km and less than 100m above sea level (Small and Nicholls, 2003). The coastal zones in the vicinity of tidal inlets, which are commonly utilized for navigation, sand mining, waterfront developments, fishing and recreation, are under particularly high population pressure. The intensive population concentration and excessive natural resources exploitation in these areas could lead to biodiversity loss, destruction of habitats, pollution, as well as conflicts between potential uses, and space congestion problems, which will only be exacerbated by foreshadowed climate change. In the case of tidal inlets, the adjacent coastal zones will be affected not only by CC driven variations in oceanic processes (e.g. Sea level rise, waves), but also by CC driven variations in terrestrial processes (e.g. rainfall/runoff) (Ranasinghe et al., 2013). Any negative impacts of CC on inlet environment are therefore very likely to result in large socio-economic impacts.

Tidal inlets which connect an estuary/lagoon/river to the coast are commonly found throughout the world. While the total number of inlets around the world is to date unquantified, it is likely to be several tens of thousands (Carter and Woodroffe, 1994). Bruun and Gerritsen (1960) distinguish three inlet classes based on their origin, as geological origin (also known as drowned river valleys);

littoral origin such as openings through barrier islands, and hydrological origin where a river enters the sea (directly or via an estuary/lagoon) (Figure 2.1).

Figure 2.1. Examples of the three main types of tidal inlets: (a) Golden Gate, California, USA (Geological origin or drowned river valley inlet); (b) Drum Inlet, North Carolina, USA (Littoral origin or barrier island inlet) ; (c) Maha Oya river inlet, Sri Lanka (Hydrological origin or bar-built/barrier estuary inlet) (sources: Google and Google earth images).

Geological origin inlets (e.g. The Golden Gate inlet, California, USA; Botany Bay inlet, Sydney, Australia) are believed to have been formed by glacier-fed rivers scouring through bedrock on their way to the ocean during the last Ice Age, when sea level was several hundred meters lower. Sea level rise during the Holocene has resulted in the associated river valleys being slowly drowned, forming large estuaries and wide, stable inlets.

Barrier island coasts are reported to comprise about 15% of the world's coastlines (FitzGerald et al., 2008). These coasts are formed by groups or chains of barrier islands and inlets (i.e. gaps in the island chain) that mostly occur parallel to the mainland coast. Barrier islands are formed due to the combined action of waves, winds and longshore current that result in the formation of thin strips of land that are several meters above MSL. Barrier island coasts are usually separated from the mainland by lagoons, marshes and/or tidal flats. While this type of inlet systems is found in every continent except Antarctica, a vast majority is located along the US East coast and the Gulf of Mexico, East and West coast of Alaska and East coast of South America (FitzGerald et al., 2008).

The third type of tidal inlets connects relatively small estuaries/lagoons, known as "bar-built" or "barrier" estuaries, to the ocean (hereafter referred to as Small Tidal Inlets or STIs for convenience). These are commonly found in wave-dominated, microtidal mainland (Ranasinghe et al., 2013). Due to their pre-dominant occurrence in tropical and sub-tropical zones, these systems usually experience a strong seasonal cycle in system forcing comprising high and low riverflow/wave energy seasons (monsoon/non-monsoon or winter/summer). In some cases, wave direction may also have a strong annual signal, particularly in monsoonal areas. As mentioned in Chapter 1, due to their high vulnerability to CC impacts, understanding and quantifying CC impacts on the stability of STIs is crucial. As a first step towards achieving that challenging goal, this introductory Chapter aims to: (a) summarise potential CC impacts on the stability of STIs, (b) conceptualise means by which the CC impacts maybe quantified using existing modelling tools, and (c) propose ways forward to enable better quantification of CC impacts on STIs.

This Chapter is structured as follows. First a brief review of the stability of STIs is provided in Section 2.2, followed by a summary of the CC processes (and their global projections) that may affect STI stability in Section 2.3. Subsequently, in Section 2.4, the quantification of CC impacts on STIs using currently available numerical modelling tools is discussed and two different modelling frameworks for data rich and data poor environments are presented. Section 2.5 provides a discussion on the inherent weaknesses in the proposed modelling frameworks and potential solutions. Finally, an overall summary and conclusions are given in Section 2.6.

2.2   Stability of Small Tidal Inlets

STI behaviour is governed by the delicate balance of oceanic processes such as tides, waves and mean sea level (MSL), and fluvial/estuarine processes such as riverflow, all of which can be significantly affected by CC. Potential impacts include (but not limited to) erosion of open coast beaches adjacent to the inlet and/or estuary margin shorelines, permanent or frequent inundation of low lying areas on estuary margins, eutrophication, and toxic algal blooms etc. Importantly, CC driven changes in forcing may affect the stability of the inlet itself, which is the main focus of this Thesis.

Inlet stability can refer to either locational stability or channel cross-sectional stability. Inlets that are cross-sectionally stable are those in which the inlet dimensions will remain more or less constant over time. Inlets that are locationally stable generally stay fixed in place over time. A locationally stable inlet may be cross-sectionally stable or unstable (e.g. intermittently closing inlets). A cross-sectionally stable inlet may also be locationally stable or unstable (e.g. alongshore migrating inlets). Inlet stability is fundamentally governed by the flow through the inlet (tidal prism and riverflow) and nearshore sediment transport in the vicinity of the inlet (Bruun, 1978). For convenience, the combination of tidal prism and riverflow is referred to as tidal prism hereon.

2.2.1   STI types

Based on their general morphodynamic behaviour, STIs can be broadly characterised into 3 main sub-categories as:

- Permanently open, locationally stable inlets (Type 1)
- Permanently open, alongshore migrating inlets (Type 2)
- Seasonally/Intermittently open, locationally stable inlets (Type 3)

Locationally stable inlets do not migrate alongshore, but may stay open (i.e. locationally and cross-sectionally stable inlets - Type 1) or close intermittently/seasonally (i.e. locationally stable but cross-sectionally unstable inlets - Type 3). Inlet closure may occur due to longshore sediment transport (on drift dominated coasts) or due to onshore migration and welding of sandbars (on swash dominated coasts) (Ranasinghe at el., 1999). These two processes are conceptually described in Figure 2.2.

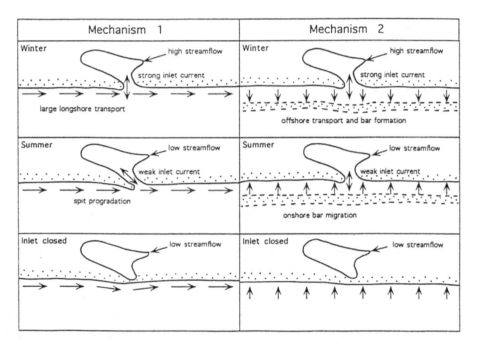

Figure 2.2. Conceptual model of inlet closure mechanisms (from Ranasinghe et al., 1999).

The main inlet morphodynamic phenomenon that characterises Type 2 STIs is alongshore migration of the inlet. The migration process of an STI (Type 2) is described in Figure 2.3 (Davis and FitzGerald, 2004). When wave-induced longshore sediment transport adds sand to the updrift side of the inlet, the inlet cross-sectional area is reduced, thus increasing flow velocities through the inlet and a greater scouring capacity in the inlet throat. As the tidal currents scour the channel by removing sand, the downdrift side of the inlet channel is eroded preferentially and the inlet migrates in the downdrift direction. In general, the migration rate depends on the magnitude of the littoral drift (sediment supply and wave energy), the ebb tidal current velocity, and on the composition of the channel bank (FitzGerald, 1988). An elongation of the inlet channel due to the inlet migration often results in a breaching of the updrift sand spit during severe storms and/or extreme riverflow events, forming a new inlet which provides a shorter, more hydraulically efficient route for tidal exchange. The new hydraulically efficient inlet will stay open while the less efficient old inlet gradually closes. Most inlets on littoral drift shores migrate in the direction of the prevailing littoral drift.

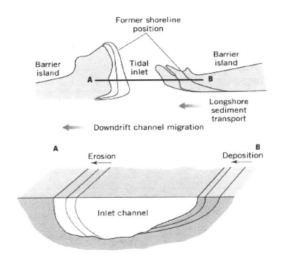

Figure 2.3. Conceptual model of inlet migration (from Davis and FitzGerald, 2004).

2.2.2   Empirical relationships to determine inlet stability

There are several empirical methods to determine the inlet stability. The most widely used empirical method is the relationship between the tidal prism and the inlet minimum cross-sectional area below mean sea level (i.e. the A-P relationship). The cross-sectional area of an inlet has been shown to be proportional to (or in equilibrium with) the volume of water flowing through it during a half tidal cycle (tidal prism) and was quantified by O'Brien (1931, 1969) as:

$A_c = a.P^n$ (2.1) where:

A_c: minimum cross sectional area of inlet gorge (m²),

P: spring tidal prism (m³),

a and n: empirical coefficients.

Subsequently, this relationship was refined by Jarrett (1976) for structured versus unstructured inlets and inlets with varying wave energy via a comprehensive investigation of inlets along Pacific Ocean, Gulf of Mexico and Atlantic Ocean Coasts of USA.

The other widely used method is the Escoffier diagram (Escoffier, 1940), which is essentially a hydraulic stability curve in which maximum flow velocity in the inlet channel is plotted against cross sectional flow area (Figure 2.4). According to this diagram, an inlet which has a cross sectional area larger than a critical flow area (point A) will be hydraulically stable.

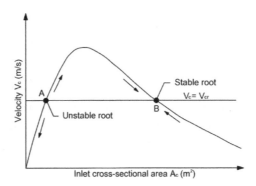

Figure 2.4. Escoffier closure diagram (after Escoffier, 1940).

The methods mentioned above only determine the inlet cross-sectional stability, but not locational stability. Bruun (1978) described the overall (both cross-sectional and locational) inlet stability criterion through the ratio:

$$r = \frac{P}{M_{tot}} \tag{2.2}$$

where: M_{tot} is the total annual littoral drift (m³/year), and P is the tidal prism (m³/tidal cycle).

According to the value of P/M_{tot}, the overall stability of an inlet is rated as good, fair, or poor as detailed in Table 2.1.

Table 2.1. The Bruun criteria for inlet stability

$r = \dfrac{P}{M_{tot}}$	**Inlet stability condition**
> 150	*Good* – predominantly tidal flow by-passers; entrance with little or no ocean bar outside gorge and good flushing
100 – 150	*Fair* – mix of bar-by-passing and flow-by-passing; entrance has low ocean bars, navigation problems usually minor
50 – 100	*Fair to poor* – inlet is typically bar-by-passing and unstable; entrance with wider and higher ocean bars, increasing navigation problems
20 - 50	*Poor* – inlet becomes unstable with non-permanent overflow channels; entrance with wide and shallow ocean bars, navigation difficult
< 20	The entrances become *unstable* "overflow channels" rather than permanent inlets.

2.3   Potential Climate Change drivers of Small Tidal Inlet stability

The stability of STIs will be affected by CC driven variations in oceanic processes (Sea level rise, wave characteristics) and also in terrestrial processes (rainfall/riverflow) as variations in one or more of these phenomena may change the flow through the inlet (tidal prism) and/or littoral transport. For example, CC driven variations in system forcing may turn a permanently open, locationally stable inlet (Type 1) into an alongshore migrating, permanently open inlet (Type 2); or, a seasonally closing, locationally stable inlet (Type 3) into a permanently open, alongshore migrating inlet (Type 2).

For the sake of completeness, presently available global projections of these CC drivers of inlet stability are briefly summarised below. In the absence of local scale projections (i.e. data poor environments), these coarse global projections may be used in first-pass CC impact assessments as described in Section 2.4.2.

Sea level rise and Relative Sea level rise

The Fifth Assessment Report AR5 of the IPCC (2013) projects that global mean sea level will continue to rise during the 21st century due to the increasing of ocean warming and mass loss from glaciers and ice sheets. Projections indicate that global mean Sea level rise (SLR) for 2081-2100 (relative to 1986-2005) will likely be in the range of 0.26 m to 0.82 m (Table 2.2) with the most pessimistic RCP8.5 scenario projecting a SLR of 0.52 m to 0.98 m, with an SLR rate of 8-16 mm/yr during 2081-2100 (Figure 2.5).

Table 2.2. Projected global mean sea level rise (m) during the 21st century relative to 1986-2005 (from IPCC 2013)

Scenarios	2046-2065		2081-2100	
	Median	*Likely* Range of mean	Median	*Likely* Range of mean
RCP2.6	0.24	0.17 to 0.32	0.40	0.26 to 0.55
RCP4.5	0.26	0.19 to 0.33	0.47	0.32 to 0.63
RCP6.0	0.25	0.18 to 0.32	0.48	0.33 to 0.63
RCP8.5	0.30	0.22 to 0.38	0.63	0.45 to 0.82

Where available and applicable, regional contributors to sea level such as meteo-oceanographic factors (e.g. ocean currents such as Gulf Steam and The East Australian Current, spatial variations in thermal expansion), changes in the regional gravity field of the Earth (i.e. higher SLR at areas farther away from areas of ice melt, and vice versa), and vertical land movements (e.g. subsidence due to gas/water extraction, uplift due to post-glacial rebound) should be added to the above global average SLR projections to derive locally applicable Relative SLR (RSLR) projections (Nicholls et al., 2014).

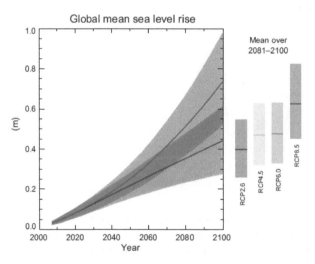

Figure 2.5. Projections of global mean sea level rise over the 21st century relative to 1986-2005. The shaded bands indicate the likely ranges. The coloured vertical bars show the likely ranges for the mean over 2081-2100 for the 4 RCP scenarios, with the horizontal lines within the vertical bars indicating the corresponding median values (from IPCC 2013).

CC driven variations in wave conditions

CC is also expected to affect the global wave climate. Hemer et al. (2013) presents the projected changes in global wave climate from a community-derived multi-model ensemble. The projected changes of wave climate include changes in significant wave height (H_s), mean wave period (T_M) and mean wave direction (θ_M) (Figures 2.6-2.8 and Table 2.3).

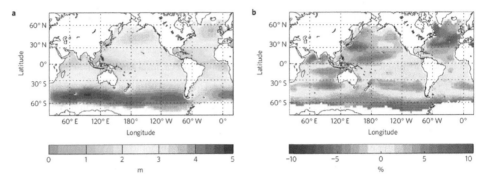

Figure 2.6. Projected future changes in significant wave height. (a) annual mean significant H_s for the present (~1979-2009). (b) projected changes in annual mean H_s for the future (~2070-2100) relative to the present (~1979-2009) (% change) (from Hemer et al., 2013).

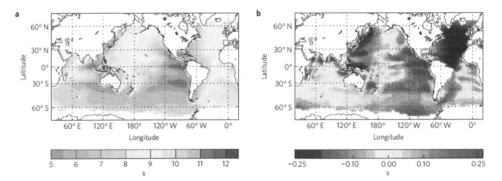

Figure 2.7. Projected future changes in mean wave period. (a) annual mean T_M for the present (~1979-2009). (b) projected changes in annual mean T_M for the future (~2070-2100) relative to the present (~1979-2009) (absolute change (s)) (from Hemer et al., 2013).

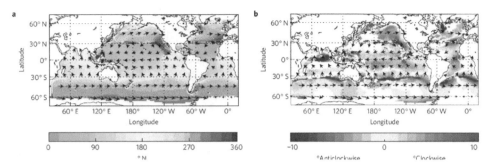

Figure 2.8. Projected changes in mean wave direction. (a) annual mean wave direction θ_M (degrees clockwise from North) for the present (~1979-2009). (b) projected changes in annual mean wave direction θ_M for the future (~2070-2100) relative to the present (~1979-2009) (absolute change, degrees clockwise). Vectors in (b) indicate the directions in the left colour bar. Colours indicate the magnitude of projected change following to the right colour bar (from Hemer et al., 2013).

More than 25.8% of the total global ocean area is projected to decrease in annual mean H_s, while an increase in H_s is projected for only about 7.1% of the total area. More than 30% of the global ocean is projected to (marginally) increase in annual mean T_M, while clockwise and anti-clockwise rotations in wave direction (θ_M) are projected for about 40% of the global ocean.

Table 2.3. Percentage areas of global ocean where robust changes in significant wave height, mean wave period, and mean wave direction are projected (after Hemer et al., 2013).

Annual values	Percentage area of robust projected *increase*	Percentage area of robust projected *decrease*
H_S	7.1	25.8
T_M	30.2	19
θ_M	18.4 *(clockwise)*	19.7 *(anti-clockwise)*

CC driven variations in riverflow

Precipitation and temperature changes lead to the changes in runoff and availability of water. IPCC AR5 global projections for the most extreme scenario RCP8.5, indicate greater than 30% decreases in annual runoff in parts of southern Europe, the Middle East and southern Africa, and similar percentage increases in the high northern latitudes by the end of the 21st century relative to the present (Figure 2.9).

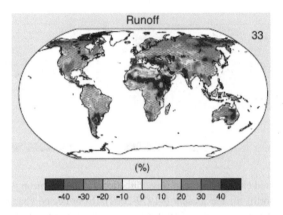

Figure 2.9. Projected change in annual mean runoff by 2081-2100 relative to 1986-2005 for the worst case RCP8.5 scenario. Regions where the multi-model mean change is less than one standard deviation of internal variability is shown by hatching. Regions where the multi-model mean change is greater than two standard deviations of internal variability and where 90% of models agree on the sign of change is shown by stippling (from IPCC 2013).

2.4   Quantifying Climate Change impacts on the stability of Small Tidal Inlets

Ideally, to investigate CC impacts on inlet systems, a coastal area process based morphodynamic model simulation needs to be undertaken for the entire period of interest, typically 50-100 years. Such a model would include all the effects of episodic events (e.g. storms, surges, extreme riverflow events), medium-term phenomena (e.g. changes in average wave height/direction and annual riverflow), and long-term CC effects (e.g. (R)SLR). However, attempts to undertake coastal area morphodynamic simulations (with concurrent tidal and wave forcing) exceeding a few years, even with reasonably reduced forcing conditions, have been unsuccessful to date (Lesser, 2009). Numerous attempts to overcome this problem have been made since the 1990s using very different approaches (De Vriend et al., 1993; Dabees and Kamphuis, 2000; Hanson et al., 2003; Roelvink, 2006), but all have only been partially successful. The main problem being the accumulation of numerical errors in the computational domain eventually resulting in morphological instabilities, especially with irregular wave forcing. While the use of morphological upscaling methods such as the MORFAC (Roelvink, 2006; Ranasinghe et al., 2011) to speed up the morphodynamic evolution, have allowed longer morphodynamic simulations, the disadvantage of such approaches seems to be that any initially small instabilities will also grow multiplicatively in time, eventually leading to unreliable model predictions. Even with the MORFAC approach, long term morphodynamic simulations have only been successfully undertaken for tide dominated estuaries and inlets (van der Wegen and Roelvink, 2008; Dastgheib et al., 2008; van der Wegen et al., 2011; van der Wegen, 2013), but not for wave dominated inlets with the complexity of irregular waves (let alone seasonally changing riverflows). To date, morphodynamic simulations of wave-dominated inlets that have resulted in realistic results have been limited in duration to a few months (Ranasinghe et al., 1999; Bertin et al., 2009; Bruneau et al., 2011). Thus, a process based coastal area (2DH) model that is capable of producing robust 50-100 year morphodynamic predictions with concurrent water level, wave and riverflow forcing does not currently exist. Even if such a model were available, the high computational demand of the model will not allow the multiple simulations that would be required to adequately account for the large uncertainty stemming from multiple sources (e.g. GHG scenarios, GCMs, morphodynamic model) that are inherent to CC impact studies.

An alternative approach is to undertake strategic 'snap-shot' simulations using process based coastal area models to gain some qualitative insights on how CC may affect inlet stability. In this approach, a model that has been validated for present conditions can be applied with future forcing for a simulation length of about 1 year at the desired future times (e.g. 2050, 2100) such that the annual cycle of forcing and/or morphological behaviour is represented. This approach will provide a good qualitative assessment of the potential impact of CC on inlet stability.

Snap-shot simulations to assess CC impacts on the stability of STIs may be designed and implemented in two different ways depending on whether the application is in a "data rich" or "data poor" environment. The basic rationale of both approaches is to 'validate' a 2DH morphodynamic model to first reproduce the main observed contemporary system morphodynamic characteristics (e.g. seasonal/intermittently closure; permanently open state; or alongshore migration) and then to use the validated model to obtain projections of system behaviour under climate change forcing. The two approaches are described below.

2.4.1   Data rich environments

A study area can be considered as 'data rich' when detailed bathymetries of the estuary, inlet and nearshore zone (extending to about 20m depth); at least a few years of wave, wind and riverflow data (ideally exceeding 10 years to encapsulate inter-annual variability); and downscaled future CC modified wave and riverflow data are available for the desired planning horizon (e.g. 2100).

CC impact assessments invariably contain large uncertainties. A sequentially applied train of numerical models could be used to account for these uncertainties. Ruessink and Ranasinghe (2014) present an ensemble modelling framework (Figure 2.10) that could be used in a data rich environment for robust assessment of CC impacts.

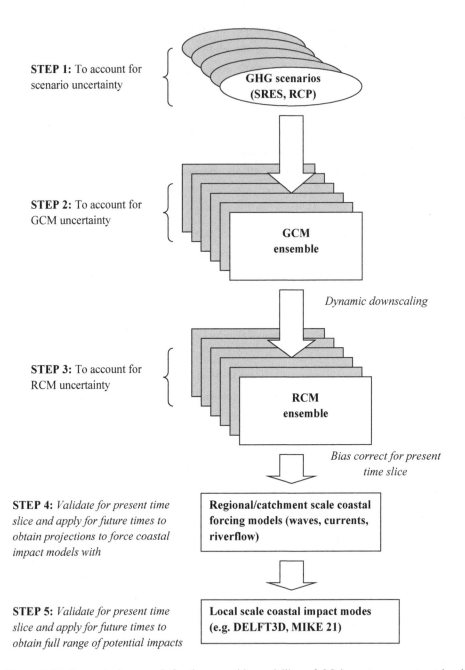

STEP 1: To account for scenario uncertainty

STEP 2: To account for GCM uncertainty

STEP 3: To account for RCM uncertainty

STEP 4: *Validate for present time slice and apply for future times to obtain projections to force coastal impact models with*

STEP 5: *Validate for present time slice and apply for future times to obtain full range of potential impacts*

Figure 2.10. Conceptual approach for the ensemble modelling of CC impacts on coasts at local scale (after Ruessink and Ranasinghe, 2014).

The approach starts from a global scale and zooms into a local site scale (~10 km length scale) via a logical sequencing of Global Climate Models (GCMs), Regional Climate Models (RCMs),

Regional wave/hydrodynamic/catchment models, local wave models, and coastal impact models. This ensemble modelling approach will provide a number of different projections of the coastal impact under investigation. The range of projections will account for GHG scenario uncertainty, GCM uncertainty, and RCM uncertainty. If necessary regional/catchment scale forcing model and coastal impact model uncertainty can also be included in this approach at a significant computing cost. The range of coastal impact projections thus obtained can then be statistically analysed to obtain not only a best estimate of coastal impacts but also the range of uncertainty associated with the projections, which will enable coastal managers/planners to make risk informed decisions.

In Step 5 of the above approach, a coastal impact model appropriate for investigating the desired system diagnostic needs to be adopted. In the case considered herein, i.e. the stability of STIs, a 2DH morphodynamic model such as *Delft3D or Mike21* may be used as described below (see also Figure 2.11).

Hydrodynamic calibration/validation

As a necessary first step, the process based model should be calibrated/validated against hydrodynamic measurements, such as water level and velocity, at several locations within the inlet-estuary system. Ideally the model should be calibrated against data for at least one full spring-neap cycle during both high and low riverflow conditions and subsequently validated for two spring-neap cycles (both high and low riverflow conditions). To achieve a good model skill, it is crucial that the grid size at the inlet channel and in the surf zone along the adjacent coast is sufficiently fine to resolve physical processes occurring therein and that offshore and lateral domain boundaries are sufficiently far from the inlet to prevent any boundary effects from propagating into the vicinity of the inlet.

Morphodynamic validation

The hydrodynamically validated model may then be used to simulate the present morphodynamic behaviour of the STI. The target of this 'present simulation' (PS) is to gain confidence in the model's ability to simulate system morphodynamics by reproducing the general contemporary morphodynamic behaviour of the system (e.g. closed/open, locationally stable/migrating). Simulations should span at least one year to represent the annual cycle of riverflow (high/low seasons) and wave conditions (winter/summer or monsoon/non-monsoon), or in the case of seasonally/intermittently closing inlets, until inlet closure occurs. As the ensemble approach recommended in Figure 2.10 will necessitate a substantial number of morphodynamic simulations, undertaking all simulations with high resolution time series forcing will constitute an almost impracticable computational effort. Therefore, some level of temporal aggregation will be required. In most cases, it will most likely be sufficient to use spring-neap cycle averaged riverflow and wave conditions in a simulation of two full spring-neap cycles (~28 days) together with a morphological acceleration (MORFAC) of 13 to allow the representation of approximately one year of morphodynamics.

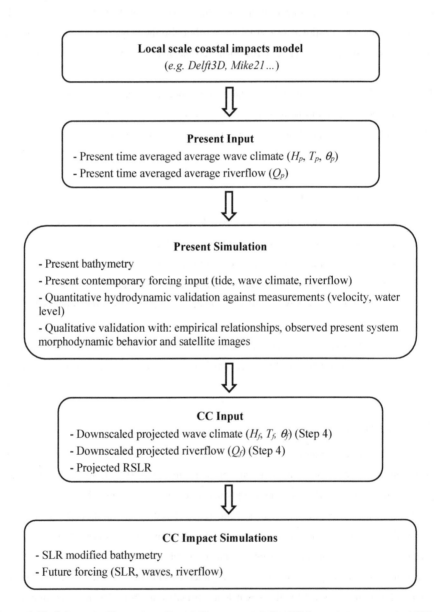

Figure 2.11. Schematic illustration of modelling approach for CC impacts assessment at STIs in data rich environments. Subscripts 'p' and 'f' refer to 'present' and 'future' respectively.

STI behaviour can strongly depend on the co-occurrence characteristics of riverflow and wave conditions. For example, when high energy (or highly obliquely incident) waves (i.e. large longshore sediment transport - LST) co-exist with high riverflow (i.e. high ebb flow velocities through inlet), the hydraulic capacity to flush out the sediment deposited in the inlet mouth will be high, and therefore the inlet will remain stable. On the other hand, when riverflow is low (with the same wave conditions), the ebb flow velocities will be lower, and therefore the hydraulic capacity to flush out sand deposited in the inlet mouth will be much reduced, potentially leading to inlet closure or migration. Thus, it is important that any input reductions made to increase computational efficiency do not affect concurrent temporal variations in forcing that exist in nature.

Model results may be compared with empirical relationships such as the A-P relationship, Escoffier curve, and Bruun inlet stability criteria. Furthermore, model results may also be qualitatively validated using any available aerial/satellite images of the study areas.

Climate Change impact projections

The validated model may now be applied to investigate CC impacts on the STI. The CC impact simulations should account for the full range of potential CC driven variations in system forcing (mean water level, waves, riverflow). The future projected RSLR (accounting also for regional effects on sea level) may be calculated using the approach prescribed by Nicholls et al. (2014), while CC modified future wave climate and riverflow can be obtained from the dynamically downscaled, regional output from Regional/catchment scale coastal model, after Step 4 in the Ensemble modelling approach shown in Figure 2.10.

The CC impact snap-shot simulations may also be undertaken for the same duration as the PS (unless in the case of Seasonally/Intermittently open inlets, in which the simulation needs to continue only until inlet closure occurs). An important long term process that needs to be (and can be) accounted for in these simulations is SLR driven basin infilling. This process occurs due to the increase in estuary/lagoon (or basin) volume below mean water level as a result of SLR (i.e. 'accommodation space'). As the basin strives to maintain an equilibrium volume, it will import sediment from offshore to raise the basin bed level such that the pre-SLR basin volume is maintained. Depending on sediment availability, when a sand volume equal to the SLR induced accommodation space (SLR x surface area of basin) is imported into the basin, the basin will reach equilibrium. Stive et al. (1998), however point out that in most situations there will be a lag between the rate of SLR and basin infilling due to the time scale disparity between hydrodynamic forcing and morphological response. Using volume balance considerations, Ranasinghe et al. (2013) have shown that for STIs, this lag could be about 0.5 over the 21st century (i.e. the basin infill volume by the end of the 21st century is equal to half of the SLR driven increase in accommodation space over the same period). Using these approximations, the long term process of basin infilling can be accommodated in snap-shot simulations by adjusting the initial bathymetry of the future simulations. Attention should be paid however to ensure that the adjusted future bathymetry maintains the contemporary basin hypsometry (Boon and Byrne, 1981).

A series of CC impact snap-shot simulations, where individual CC modified forcings are sequentially excluded from an initial all-inclusive simulation (e.g. SLR + CC Waves + CC Riverflow; SLR + CC Waves; SLR + CC Riverflow; and SLR only) should ideally be undertaken to investigate the relative contribution of the various CC modified forcings to potential changes in inlet stability.

2.4.2   Data poor environments

At most locations however, especially in developing countries, the data required for the approach described in Section 2.4.1 is not available. Especially, good bathymetry data and future downscaled forcing from sophisticated and computationally expensive Regional Scale modelling (Step 3 in Figure 2.10) approach are very unlikely to be available at most locations. In such data poor environments, a strategic schematized modelling approach may be used as described below to obtain a 'first-pass-assessment' of CC impacts on STIs (see also Figure 2.12). This approach does however assume the availability of at least good guestimates of contemporary monthly averaged riverflows and wave conditions.

Schematized bathymetry
In this approach, a simplified schematized bathymetry can be used to qualitatively represent the real system bathymetry. Over the last decade, this approach has been successfully used to gain qualitative insights into inlet morphodynamics of diverse inlet-estuary/lagoon systems at various time scales (Marciano et al., 2005; Dastgheib et al., 2008; van der Wegen and Roelvink, 2008; van der Wegen et al., 2010; Nahon et al., 2012; Dissanayake et al., 2012; van Maanen et al., 2013; Zhou et al., 2014).

> **Local scale coastal impacts model**
> *(e.g. Delft3D, Mike21...)*

> **Present Input**
> - Present monthly average wave climate (H_p, T_p, θ_p)
> - Present monthly average riverflow (Q_p)

> **Present Simulation**
> - Schematized initial bathymetry
> - Present contemporary forcing input (tide, wave climate, riverflow)
> - Qualitative validation with: empirical relationships, observed present system morphodynamic behavior and satellite images (if available)

> **CC Input**
> - Published projected changed wave climate (H_f, T_f, θ_f) (Hemer et al., 2013)
> - Published projected riverflow (Q_f) (IPCC, 2013)
> - Projected SLR (IPCC, 2013)

> **CC Impact Simulations**
> - SLR modified bathymetry
> - Future forcing (SLR, waves, riverflow), account for all possible combinations of CC modified forcing

Figure 2.12. Schematic illustration of modelling approach for CC impact assessment at STIs in data poor environments. Subscripts 'p' and 'f' refer to 'present' and 'future' respectively.

Following the philosophy adopted in these previous studies, an STI system schematized bathymetry could consist of a rectangular estuary/lagoon of constant depth connected to the ocean via a straight, constant depth channel. The dimensions of the schematized system (estuary/lagoon and inlet channel width/length) should be chosen such that they closely represent a real-life system, based on for e.g. Google Earth images. The mean depth of the estuary/inlet channel may be gleaned from any available literature, from local expert judgment or one-off, low-tech spot measurements.

The bathymetry of the ocean side could consist of shore-parallel depth contours such that a Dean's equilibrium profile (with D_{50} depending on local conditions) is followed up to about 10-20m depth. Riverflow may be introduced into the estuary/lagoon at roughly the same location the main riverflow enters the system based on Google Earth or Satellite images. Care should be taken to ensure that the grid size at the small inlet mouth and in the surf zone along the adjacent coast is fine enough to adequately resolve the physical processes occurring therein. Figure 2.13 shows an example initial bathymetry of such a schematized STI system.

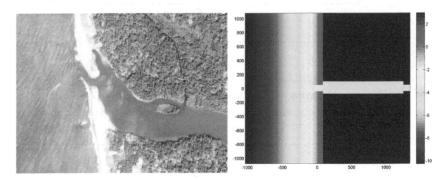

Figure 2.13. Example of schematized bathymetry for Maha Oya Inlet (Right) with: inlet width = 100m, inlet length = 70 m, inlet depth = 3 m; estuary width = 200 m, estuary length = 1000 m, estuary depth = 3.5 m; river width = 100 m, nearshore bed slope follows Dean equilibrium profile for D_{50} of 0.25 mm. Satellite image (left) from Google Earth.

CC modified forcing

To circumnavigate the typical unavailability of downscaled future forcing in data poor environments, future forcing needs to be sourced from the freely available sources described in Section 2.3 above. Future SLR may be obtained from Table 2.2 and Figure 2.5 while future riverflows may be obtained from Figure 2.9 for the worst case IPCC scenario. Future wave forcing may be obtained from the ensemble global downscaling results presented by Hemer et al. (2013) (reproduced above in Figs 2.6-2.8). It should be borne in mind that these freely available global scale future projections, particularly of riverflow and waves, are at a much coarser resolution than that would be produced by a site-specific downscaling study (as in data rich environments). Therefore, the aforementioned global maps of projected change will only give approximate indications of how these system forcings might change (e.g. as area averaged % increase/decrease relative to the present).These changes then need to be superimposed on available contemporary riverflow/wave data to derive future forcing conditions for the coastal impact model.

Morphodynamic validation

As in the case of data rich environments, a Present simulation (PS) should be undertaken to at least qualitatively validate the model against observed present behaviour of the STI. The model, initialised with the above described schematized bathymetry can be forced with schematized harmonic tidal forcing (say, an M2 harmonic with approximate observed mean tidal amplitude in the study area) and monthly averaged time series of waves and of riverflow.

In this case too, the PS should span at least one year to represent the annual cycles of riverflow and wave conditions, or in the case of seasonally/intermittently closing inlets, until inlet closure occurs. Due to the monthly averaged forcing, it is sufficient here to use a MORFAC of 30, such that one day of the hydrodynamic simulation will represent a month (30 days) of morphodynamics. Model results can be qualitatively validated following the same approach described for PS validation in data rich environments (i.e. using empirical relationships and aerial/satellite imagery).

Climate Change impact projections

The qualitatively validated model may then be forced with CC modified forcing conditions (derived as described above), following the same approach outlined for data rich environments. Due to the flat bed of the initial bathymetry, here, the basin infilling effect may simply be represented by raising the estuary/lagoon bed level by approximately half the SLR amount following the argumentation presented by Ranasinghe et al. (2013).

While this approach will provide a useful first-pass assessment of CC impacts on the stability of STIs in data poor environments, the uncertainty associated with the projections will be high due to the coarse and approximate model forcing and schematisation of system bathymetry. Therefore, if this approach indicates any signs of future inlet stability being significantly different from the present situation, it would be prudent to undertake a more detailed study (including targeted data collection and GCM downscaling).

2.5   Discussion

While the snap-shot simulation approach described above will provide insights on inlet stability that are useful for coastal zone management/planning, it also has several shortcomings. One of the major shortcomings is that this approach will not be able to take into account slow gradual morphological changes (excepting the SLR driven basin infilling process) that may occur from 'present' to 'future'. For example, any gradual CC driven changes in average wave direction could change the orientation of the coastline along which the target STI is located. This may have implications on the future longshore sediment transport rate and inlet dimensions (and therefore tidal prism), thus affecting inlet stability. Slow changes in longshore sediment transport rate and riverflow may also in some cases result in the development of extensive flood shoals in the lower reaches of the estuary/lagoon (close to the inlet channel) which could affect tidal attenuation, and hence the tidal prism, thus influencing inlet stability. Furthermore, the above snap-shot modelling approach will not account for any irreversible morphological changes that may be caused by CC modified extreme storm surge events (e.g. breaching of new inlets) and riverflows (e.g. ebb/flood delta development).

To obtain reliable projections of CC impacts on the stability of STIs, what is ideally required is a multi-scale coastal impact model that can accurately simulate the various physical processes occurring at different spatio-temporal scales. Such a model should incorporate both cross-shore (vertically non-uniform) and longshore (more or less vertically uniform) hydrodynamics to simulate coastal hydrodynamics relevant for episodic, medium-term, and long-term STI morpodynamics. Thus, the model needs to be a coastal area model with at least quasi 3D if not fully 3D hydrodynamics. To avoid the inevitable far field instabilities that will creep into the area of interest when using a gridded approach for morphodynamics, some spatio-temporal aggregation of hydrodyamics prior to calculating bed level changes may be required. However, as such a new multi-scale modelling approach might take years (or decades) to develop, an interim solution may be found in scale aggregated (or reduced complexity) models. This type of models, if correctly developed, may be used to obtain the full temporal evolution of CC driven variations in STI stability. In such a scale aggregated model, following the rationale presented by Stive and Wang (2003) and Ranasinghe et al. (2013), the main physical processes governing the main diagnostic (in this case: inlet stability) may be parameterised and collectively represented by a fully explicit governing equation. This would result in a very fast model that enables the multiple simulations required to quantify the uncertainties associated with assessing CC impacts on STI stability.

2.6   Summary and Conclusions

Climate change driven variations in mean water level (i.e. SLR), wave conditions and riverflow are likely to affect the stability of the thousands of Small tidal Inlets (STIs, or bar-built/barrier estuary systems) around the world. Due to their pre-dominant occurrence in tropical and sub-tropical zones, this type of inlets are commonly found in developing countries, where data availability is generally sub-optimal (i.e. data poor environments) and community resilience to coastal changes is low. Furthermore, STI environs in developing countries especially host a number of economic activities (and thousands of associated livelihoods) such as harbouring sea going fishing vessels, inland fisheries (e.g. prawn farming), tourist hotels and tourism associated recreational activities which contribute significantly to the national GDPs. This combination of pre-dominant occurrence in developing countries, socio-economic relevance and low community resilience, general lack of data, and high sensitivity to seasonal forcing makes STIs potentially very vulnerable to CC impacts.

This Chapter provides a summary description of how CC may affect the stability of STIs and how these CC impacts maybe quantified using existing modelling tools. Due to the unavailability of process based models that can be confidently applied with concurrent time varying water level, wave and riverflow forcing over typical CC impact assessment time scales (~100 years), 'snap-shot' simulations (~1 year duration) of process based coastal area morphodyamic models are proposed as a means of obtaining qualitative assessments of CC impacts on STIs. Two different 'snap-shot' modelling frameworks for data rich and data poor environments are presented. The main shortcomings of the proposed 'snap-shot' modelling approach are identified as non-consideration of CC driven slow gradual morphological changes (except SLR driven basin infilling) and irreversible morphological changes due to CC modified extreme storm surge, wave, riverflow events. To obtain more reliable assessment of CC impacts on STIs, the development of process based multi-scale coastal area morphodynamic models and scale aggregated morphodynamic models are identified as future research needs.

References

Bertin, X., Fortunato, A.B., Oliveira, A., 2009. A modeling-based analysis of processes driving wave-dominated inlets. Continental Shelf Research, 29 (5-6), 819-834.

Boon, J.D., Byrne, R.J., 1981. On basin hypsometry and the morphodynamic response of coastal inlet systems. Marine Geology, 40, 27-48.

Bruneau, N., Fortunato, A.B., Dodet, G., Freire, P., Oliveira, A., Bertin, X., 2011. Future evolution of a tidal inlet due to changes in wave climate, sea level and lagoon morphology: O´bidos lagoon, Portugal. Continental Shelf Research 31, 1915-1930.

Bruun, P., 1978. Stability of tidal inlets – theory and engineering. Developments in Geotechnical Engineering. Elsevier Scientific, Amsterdam, 510p.

Bruun, P., Gerritsen, F., 1960. Stability of coastal inlets. North-Holland Publishing Co., Amsterdam, 123p.

Carter, R. W. G. and Woodroffe, C. D., 1994. Coastal evolution: Late quaternary shoreline morphodynamics. Cambridge university press, 517p.

Dabees, M., Kamphuis, J.W., 2000. ONELINE: Efficient Modeling of 3D Beach Change. 2000. Proceedings of the 27th International Conference on Coastal Engineering, Sydney, Australia, ASCE, pp. 2700-2713.

Dastgheib, A., Roelvink, J.A., Wang, Z.B., 2008. Long-term process-based morphological modeling of the Marsdiep Tidal Basin. Marine Geology, 256 (1-4), 90-100.

Davis, R.A., FitzGerald, D.M., 2004. Beaches and coasts. Blackwell Publishing. 419p.

DeVriend, H.J., Zyserman, J., Nicholson, J., Roelvink. J.A., Pechon, P., Southgate, H.N., 1993. Medium term 2DH coastal area modelling. Coastal Engineering, 21, 193-224.

Dissanayake P. K.., Ranasinghe, R., Roelvink, D., 2012. The morphological response of large tidal inlet/basin systems to sea level rise. Climatic Change, 113, 253-276.

Escoffier, F.F., 1940. The stability of tidal inlets. Shore and Beach, 8,111-114.

FitzGerald, D.M., 1988. Shoreline erosional-depositional processes associated with tidal inlets. In: Lecture Notes on Coastal and Estuarine Studies, Vol.129, Hydrodynamics and Sediment Dynamics of Tidal Inlets (Eds. Aubrey, D.G., Weishar, L.), Springer-Verlag, pp. 186-225.

FitzGerald, D.M., Fenster, M.S., Argow, B.A., Buynevich, I.V., 2008. Coastal impacts due to sea-level rise. Annual Review of Earth and Planetary Sciences, 36, 601-647.

Hanson, H., Aarninkhof, S., Capobianco, M., Jimenez, J.A., Larson, M., Nicholls, R., Plant, N., Southgate, H.N., Steetzel, H.J., Stive, M.J.F., De Vriend, H.J., 2003. Modelling coastal evolution on early to decadal time scales. Journal of Coastal Research, 19(4), 790-811.

Hemer, M., Fan, Y., Mori, N., Semedo, A., Wang, X.L., 2013. Projected changes in wave climate from a multi-model ensemble. Nature Climate Change, 3, 471-476.

IPCC, 2013. Summary for Policymakers. In: Climate Change 2013: The Physical Science Basis. Contribution of Working Group I to the Fifth Assessment Report of the Intergovernmental Panel on Climate Change [Stocker, T.F., D. Qin, G.-K. Plattner, M. Tignor, S.K. Allen, J. Boschung, A. Nauels, Y. Xia, V. Bex and P.M. Midgley (eds.)]. Cambridge University Press, Cambridge, United Kingdom and New York, NY, USA.

Jarrett, J.T., 1976. Tidal prism – inlet area relationships. Technical Report GITI No.3, CERC, U.S. Army Engineer Waterways Experiment Station, Vicksburg, MS.

Lesser, G., 2009. An approach to medium-term coastal morphological modeling. PhD thesis. UNESCO-IHE/Delft University of Technology, ISBN 978-0-415-55668-2.

Marciano, R., Wang, Z.B., Hibma, A., de Vriend, H.J., Defina, A., 2005. Modeling of channel patterns in short tidal basins. Journal of Geophysical Research, 110, F01001, doi: 10.1029/2003JF000092.

Nahon, A., Bertin, X., Fortunato, A.B., Oliveira, A., 2012. Process-based 2DH morphodynamic modeling of tidal inlets: A comparison with empirical classifications and theories. Marine Geology, 291–294, 1–11, doi:10.1016/j.margeo.2011.10.001.

Nicholls, R.J., Hanson, S., Lowe, J.A., Warrick, R.A., Lu, X., Long, A.J., 2014. Sea-level scenarios for evaluating coastal impacts. Wiley Interdisciplinary Reviews: Climate Change 5 (1), 129-150.

O'Brien, M.P., 1931. Estuary and tidal prisms related to entrance areas. Civil Engineering 1(8), 738-739.

O'Brien, M.P., 1969. Equilibrium flow areas of inlets on sandy coasts. Journal of Waterways and Harbour Division, 95, 43-52.

Ranasinghe, R., Pattiaratchi, C., Masselink, G., 1999. A morphodynamic model to simulate the seasonal closure of tidal inlets. Coastal Engineering, 37(1), 1-36.

Ranasinghe, R., Swinkels, C., Luijendijk, A., Roelvink, D., Bosboom, J., Stive, M., Walstra, D., 2011. Morphodynamic upscaling with the MORFAC approach: Dependencies and sensitivities. Coastal Engineering, 58, 806-811.

Ranasinghe, R., Duong, T.M., Uhlenbrook, S., Roelvink, D., Stive, M., 2013. Climate change impact assessment for inlet-interrupted coastlines. Nature Climate Change, 3, 83-87, DOI.10.1038/NCLIMATE1664.

Ruessink, B.G., Ranasinghe, R., 2014. Beaches. In: Coastal Environments and Global Change (Eds. Masselink, G. and Gehrels, R.). Wiley, pp. 149-176.

Roelvink, J.A., 2006. Coastal morphodynamic evolution techniques. Coastal Engineering, 53, 277-287.

Small, C., Nicholls, R.J., 2003. A global analysis of human settlement in coastal zones. Journal of Coastal Research, 19(3), 584-599.

Stive, M. J. F., Capobianco, M., Wang, Z. B., Ruol, P., Buijsman, M. C., 1998. Morphodynamics of a tidal lagoon and the adjacent coast. Proceedings of the Eighth International Biennial Conference on Physics of Estuaries and Coastal Seas, The Hague, pp.397- 407.

Stive, M.J.F., Wang, Z.B., 2003. Morphodynamic modelling of tidal basins and coastal inlets. Advances in Coastal Modelling (Ed. C. Lakhan). Elsevier Science B.V, pp. 367-392.

van der Wegen, M., 2013. Numerical modeling of the impact of sea level rise on tidal basin morphodynamics, Journal of Geophysical Research, 118, doi:10.1002/jgrf.20034.

van der Wegen, M., Jaffe, B. E., Roelvink, J. A., 2011. Process-based, morphodynamic hindcast of decadal deposition patterns in San Pablo Bay, California, 1856-1887, Journal of Geophysical Research, 116, F02008, doi: 10.1029/2009JF001614.

van der Wegen, M., Roelvink, J.A., 2008. Long-term morphodynamic evolution of a tidal embayment using a two-dimensional, process-based model. Journal of Geophysical Research., 113, C03016, doi: 10.1029/2006JC003983.

van der Wegen, M., Dastgheib, A., Roelvink, J.A., 2010. Morphodynamic modeling of tidal channel evolution in comparison to empirical PA relationship. Coastal Engineering. 57, 827-837, doi:10.1016/j.coastaleng.2010.04.003.

van Maanen, B., Coco, G., Bryan, K.R., Friedrichs, C.T., 2013b. Modeling the morphodynamic response of tidal embayments to sea-level rise. Ocean Dynamics, 63(11-12), 1249-1262, doi:10.1007/s10236-013-0649-6.

Zhou, Z, Coco, G., Jiminez, M., Olabarrieta, M, van der Wegen, M, Townend, I. 2014. Morphodynamics of river influenced back barrier tidal basins: The role of landsacpe and hydrodynamic settings, Water Resources Research, 50, doi: 10.1002/2014WR015891.

CHAPTER 3

ASSESSING CLIMATE CHANGE IMPACTS ON THE STABILITY OF SMALL TIDAL INLETS IN DATA POOR ENVIRONMENTS

3.1   Introduction

This Chapter demonstrates the application of the process based snap-shot modelling approach for data poor environments proposed in Section 2.4.2 of Chapter 2. The approach is applied to 3 case study sites representing the 3 main STI types described above: Negombo lagoon (Type 1); Kalutara lagoon (Type 2); and Maha Oya river (Type 3); all of which are located along the SW coast of Sri Lanka (Figure 3.1).

3.2   Methodology

3.2.1   Study areas

Sri Lanka, is an island nation of about 65,610 km^2 area located in the Indian Ocean, Southeast of India (Figure 3.1). The country has a tropical monsoon climate, with 2 monsoon seasons; the Northeast (NE) monsoon from November-February and the Southwest (SW) monsoon from May-September. About one third of the total annual rainfall occurs between October and December (Zubair and Chandimala, 2006). The coastal environment of Sri Lanka is wave dominated (average offshore significant wave height of 1.12m) and micro-tidal (mean tidal range of approximately 0.5m). The SW coast of Sri Lanka generally experiences the most energetic wave conditions during the SW monsoon when offshore significant wave heights are between 1 and 2 m and mean wave direction is from the SW-W octant. The beaches around the country are composed of quartz sand with grain diameters (D$_{50}$) varying between 0.2-0.45mm.

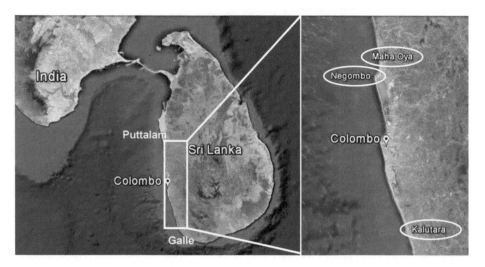

Figure 3.1. Location of Sri Lanka (left) and the 3 case study sites (right). Source: Google Earth.

Negombo lagoon

Negombo lagoon is located about 30 km North of Colombo (Figure 3.2, left) and is connected to the ocean via a permanently open, relatively wide (400 m), short (300 m) and shallow (3 m) inlet which is locationally stable. The lagoon has a surface area of around 45 km^2 and the average depth is approximately 1 m. The net annual longshore sediment transport rate in the vicinity of the inlet is insignificant (Chandramohan et al., 1990). This is due to the sheltering effect provided by the offshore reef to the west of the inlet (Figure 3.2, left). The sediment on the coast adjacent to the inlet has a D$_{50}$ of 0.25 mm (University of Moratuwa, 2003). The majority of the catchment of the Negombo lagoon is located in the SW of the country and therefore most of the riverflow into the lagoon occurs during the SW monsoon. Dandugam Oya, Ja Ela and several streams from Muthurajawela flow into the lagoon with a combined average annual riverflow volume that varies from almost 0 m^3/s during dry seasons to more than 200 m^3/s during rainy seasons (University of Moratuwa, 2003)).

Figure 3.2. Negombo lagoon (left), Kalutara lagoon (middle), Maha Oya river (right)
(Source: Google Earth).

Kalutara Lagoon

Kalutara lagoon is located about 40 km South of Colombo (Figure 3.2, middle). The Kalu River, with the second highest annual riverflow volume (7500 million m^3/yr) in the country, discharges into the ocean via this lagoon. The lagoon is connected to the ocean via a permanently open, alongshore migrating inlet with average width 150 m, length 150 m and depth 4.5m. Historically, the longshore migration of the Kalutara lagoon inlet comprised a 3-4 year cycle during which the inlet migrated about 2 km to the south (~500m/yr southerly migration) before a new, more hydraulically efficient inlet was either naturally or artificially created at the northern end of the lagoon barrier (Perera, 1993). The lagoon has a surface area of only around 2 km^2 and the average depth is approximately 3 m. The net longshore sediment transport rate in the area is about 0.5 million m^3/yr to the south (GTZ, 1994). The sediment on the coast adjacent to the inlet has a D_{50} of 0.25 mm. The catchment of the Kalu river is the 2nd largest in the country (2766 km^2) and it receives rainfall from both the SW and NE monsoons, resulting in river discharges that are consistently higher than 100 m^3/s throughout the year (with peaks exceeding 300 m^3/s in May and October). The average river discharge of the Kalu river is 250 m^3/s.

Maha Oya River

Maha Oya inlet (Figure 3.2, right), through which the Maha Oya river discharges (1571 million m^3/yr) into the ocean, is located about 40 km North of Colombo. The inlet, which is about 100 m wide, 70 m long and 3 m deep, closes whenever the riverflow is small, regardless of the prevailing longshore sediment transport regime. The lagoon surface area is about 0.2 km^2 with an average depth of 3-4 m. The net longshore sediment transport rate in the vicinity of the inlet is about 500,000m^3/year to the North (GTZ, 1994). The sediment on the coast adjacent to the inlet has a D_{50} of 0.25 mm. The 1528 km^2 catchment of Maha Oya derives most of its riverflow during the NE monsoon. The average river discharge is about 50 m^3/s with a peak of ~140 m^3/s in November.

3.2.2   Schematized bathymetries

Schematized flat bed bathymetries were created for the 3 systems based on site descriptions in published literature, Google Earth images and input from local experts. The key system dimensions thus gleaned and used in creating the schematized bathymetries (Figure 3.3) are shown in Table 3.1. The nearshore zone was configured such that it consisted shore parallel depth contours that follow a Dean's equilibrium profile in the cross-shore direction corresponding with the reported D_{50} for the area (also given in Table 3.1) down to 10m depth. Model grids for the 3 systems were generated using custom made Matlab routines which ensured that grid spacing in both cross-shore and alongshore directions were optimal for *Delft3D* computations.

Table 3.1. Key dimensions for the 3 systems used to generate the schematized bathymetries.

Dimensions	Inlet			Estuary/Lagoon				
Inlet Type	*Width (m)*	*Length (m)*	*Depth (m)*	*Width (m)*	*Length (m)*	*Depth (m)*	*Basin Area (km²)*	*D₅₀ (µm)*
Type 1 (Negombo Lagoon)	400	300	3	3500	13000	1	45	250
Type 2 (Kalutara Lagoon)	150	150	4.5	3500	500	3	1.75	250
Type 3 (Maha Oya River)	100	70	3	200	1000	3.5	0.2	250

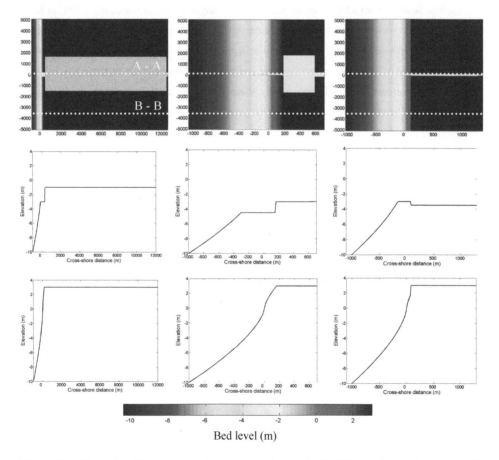

Figure 3.3. Schematized bathymetries for Negombo lagoon (left), Kalutara lagoon (middle), and Maha Oya river (right); Plan view (top); section A-A (middle); section B-B (bottom).

3.2.3   Schematized forcing

The mean ocean tidal range for all 3 systems (which are located along a single coastal stretch not exceeding 100km) was taken as 0.5m based on published values (Wijeratne, 2002; University of Moratuwa, 2003). Time series of monthly averaged riverflow were constructed for the 3 systems (Figure 3.4) based on available sparse data and local publications. Wave conditions at 10 m depth for each case study site (Figure 3.5) were extracted from regional SWAN model extending from Galle to Puttalam along the SW coast (wave directions shown have been adjusted to account for the differences between the coastline orientations at the study sites and the N-S coastline orientations adopted in the schematized bathymetries). Comparison of hourly and monthly averaged conditions at Colombo showed that monthly averaged values do not significantly under or over represent actual wave conditions. Note that while most of the waves are $< 270^0$ at Maha Oya (with respect to the coordinate system adopted in the schematized bathymetries), all waves are $> 270^0$ at Kalutara. This is due to wave refraction over a large canyon located slightly north of Kalutara. At Negombo, waves are more or less shore normal at 10 m depth. This is because all waves from the SW sector (pre-dominant offshore wave direction) are completely diffracted by an offshore low crested (and in places emergent) reef and subsequently sharply refracted to the nearshore to result in almost shore normal nearshore waves throughout the year.

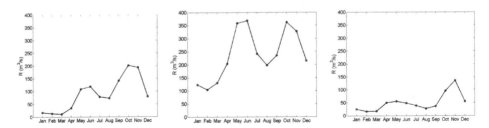

Figure 3.4. Riverflow forcing conditions for Negombo lagoon (left), Kalutara lagoon (middle), and Maha Oya river (right).

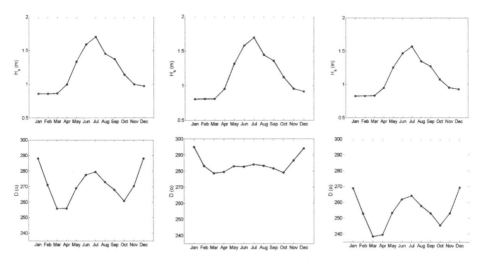

Figure 3.5. Wave forcing conditions for Negombo lagoon (left), Kalutara lagoon (middle) and Maha Oya river (right); Top: Significant wave height, Bottom: Mean wave direction.

3.2.4   Process based model

The process based model *Delft3D* was extensively used in this study. *Delft3D* seamlessly combines a short wave driver (SWAN), a 2DH flow module, a sediment transport model (Van Rijn, 1993), and a bed level update scheme that solves the 2D sediment continuity equation. To accelerate morphodynamic computations, *Delft3D* adopts the MORFAC approach (Roelvink, 2006; Ranasinghe et al., 2011). This approach exploits the fact that time scales associated with bed level changes are generally much greater than those associated with hydrodynamic forcing by essentially multiplying the bed levels computed after each hydrodynamic time step by a time varying or constant factor (MORFAC) to enable much faster computation. The significantly upscaled new bathymetry is then used in the next hydrodynamic step. The model is fully described by Lesser et al. (2004) and is therefore not described any further here. The basic model structure is shown in Figure 3.6.

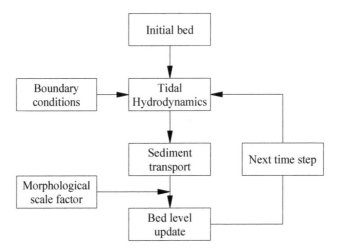

Figure 3.6. *Delft3D* Model structure.

To avoid boundary instabilities affecting the area of interest (i.e. the inlet entrance), the flow computational domains were constructed such that they extended 5 km either side of the inlet for all 3 study areas (Figure 3.7). Wave domains were created larger than flow domains to avoid any wave shadowing effect at lateral boundaries.

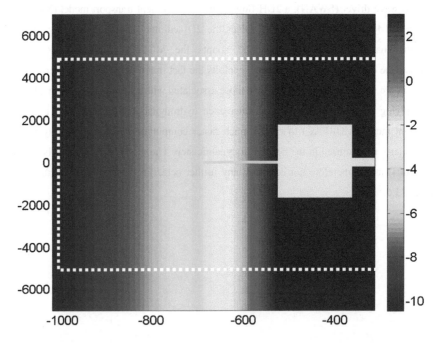

Figure 3.7. *Delft3D* flow (inside the white dashed box) and wave domains used for the Kalutara lagoon case study (Type 2 inlet).

To ensure that key physical processes in the vicinity of the inlet entrance and channel were accurately resolved by the model, high resolution (~10x10 m) grid cells were used in the (approximate) surf zone and inlet channel for all 3 study areas. Riverflow was introduced as a single (Kalutara, Maha Oya) or combined (Negombo) flow discharge at the landward boundary of the domain.

Following over 100 sensitivity tests with the 3 cases, the values shown in Table 3.2 were adopted for key model parameters in all simulations described in the ensuing sections of this Chapter.

Table 3.2. Model parameter settings.

Parameter	Adopted value
Hydrodynamic time step (s)	6
Hydrodynamic spin-up time (hrs)	12
Horizontal eddy viscosity (m^2/s)	1
Horizontal eddy diffusivity (m^2/s)	0.1
Chezy bottom friction coefficient ($m^{1/2}$s)	65
Directional wave spreading (deg)	10 (considering predominant swell conditions)
Sediment transport formula	Van Rijn (1993)
Dry cell erosion factor	0.5
Wave-flow coupling time (hrs)	1
MORFAC	30
Output interval for whole domain (hrs)	1
Output interval for pre-defined observation points and cross-sections (s)	600

To ensure truly 2DH conditions, model parameters that artificially introduce wave driven cross-shore sediment transport were not employed in the modelling undertaken in this entire study.

3.2.5   Morphodynamic model validation

For each schematized system, first a *Delft3D* simulation was undertaken with the above described contemporary forcing (i.e. 'Present simulation' - PS). Tidal forcing was introduced at the offshore boundary as a harmonic water level boundary with 0.25 m amplitude and a 12 hour period, representing the average semi-diurnal tidal condition in the case study areas. Riverflow and wave forcing followed the monthly averaged time series shown in Figures 3.4 and 3.5 above. Given the monthly time step of wave and riverflow forcing in the PS, a MORFAC of 30 was used in all simulations, thus representing (approximately) one year of morphological evolution by 12 days of hydrodynamic forcing (or 24 tidal cycles). The sensitivity of model predictions to the MORFAC value adopted was tested by re-executing the simulation with MORFACs of 15, 5 and 1 (with appropriate changes in the forcing time series and wave-flow coupling time). Only marginal differences were observed among the morphodynamic predictions of the test simulations, indicating a MORFAC of 30 could be confidently used for the simulations presented herein. Validation simulations for Type 1 and Type 2 systems were undertaken for one year to represent the annual cycle of riverflow (high/low seasons) and wave conditions (monsoon/non-monsoon periods) while the Type 3 system simulation was continued only until inlet closure occurred.

The target of this PS is to gain confidence in the model's ability to simulate system morphodynamics by reproducing the general contemporary morphodynamic behaviour (e.g. close/open, and locationally stable/migrating) of the system. To this end, model results were first qualitatively validated using available aerial/satellite images of the study areas by comparing the general observed and modelled morphodynamic behaviour of the systems. Secondly, model results were compared against several empirical relationships such as the A-P relationship (O'Brien, 1931; Jarrett 1976), Escoffier curve, and Bruun inlet stability criteria where possible. For these comparisons, it is necessary to extract information regarding inlet cross-section area (A), the tidal prism (P), maximum inlet current velocity (V_{max}), and annual longshore sediment transport rates (M). In the case studies selected here, riverflow is non-trivial, and therefore enhances the ebb tidal prism (due to the tide effect only) which is one of the two processes that govern the main diagnostic of this study, inlet stability. For convenience, therefore, all references tidal prism (P) hereon refer to the flow volume through the inlet during ebb due to the combined effect of tides and riverflow. The ways in which A, P, V_{max} and M were extracted from the model output are briefly described below.

For each case, cross-sections were pre-defined at every grid line (~10m spacing) along the inlet channel to extract and store water levels, velocities and discharges every 10 minutes (user defined output interval), which were subsequently used to calculate inlet cross-section area, tidal prism and inlet velocity. Similarly, cross-shore sections were pre-defined along the coast to calculate longshore sediment transport rates.

To calculate the A-P relation, the minimum inlet cross-section area is required. Therefore, using the 10 minute output extracted along the inlet cross-sections, the minimum inlet cross-section area A (below MWL) was determined for each tidal cycle. Tidal prism P was estimated at each cross-section by calculating the difference between consecutive cumulative discharge peaks and troughs. It should be noted that cumulative discharge is calculated internally by the model at every hydrodynamic time step (6s in this case) and then output at the pre-defined output interval (10 minutes in this case) (Note: this provides a more accurate estimate of tidal prism than when using instantaneous discharge values obtained at the output interval). As expected, the tidal prism thus calculated was spatially invariant along the inlet. The P and minimum A values per tidal cycle were then used along with appropriate C and n coefficient values for unjettied inlets specified by Jarrett (1976) to investigate inlet stability with respect to the A-P relationship.

The Escoffier curve requires the maximum flow velocity V_{max} in the inlet channel and the inlet cross-sectional flow area at the corresponding location. Comparison of maximum inlet velocities at the various pre-defined inlet cross-sections indicated that, in general, the maximum flow velocity occurred at the cross-section at the middle of the inlet channel. Therefore maximum instantaneous cross-sectionally averaged velocity per tidal cycle at the mid-inlet section was plotted against the cross-section area of the same inlet cross-section at the end of each tidal cycle to investigate inlet stability with respect to the Escoffier hydraulic stability curve.

To calculate the Bruun stability criterion, apart from the tidal prism P (which was calculated in exactly the same way as above for the A-P relationship), the ambient annual longshore sediment transport (LST) volume M is also required. In the model, the ambient LST rate will be affected by the tidal inlet as well as by the lateral boundaries. Therefore it is important that this quantity be assessed sufficiently updrift of the inlet as well as sufficiently far from the updrift boundary. To ascertain the optimal alongshore location of the cross-shore section over which M should be calculated, up to 10 cross-sections were pre-defined either side of the inlet, while ensuring that the cross-sections were long enough to always capture the full surf zone. The optimally located cross-shore section for ambient LST estimates was identified by carefully comparing the modelled LST rates across these cross-shore sections. The annual ambient LST across the optimal cross-shore section (M) was then combined with P to calculate the Bruun criterion for inlet stability $r = P/M$. This resulted in a time series of r which was averaged to obtain the annual representative r indicative of the general stability condition of the inlet.

3.2.6   Climate Change impact simulations

In each system, the validated model was then implemented via CC impact snap-shot simulations to investigate future CC impacts on the system. These simulations were also undertaken for the same duration as the PS, or until inlet closure occurred.

The forcing for the CC simulations were derived from freely available, albeit coarse resolution sources as relevant for the study areas. The adopted worst case CC driven variations in MWL (i.e.

SLR), wave conditions and riverflows are shown in Table 3.3, together with the sources from which these values were extracted.

As discussed in Section 2.4.2, due to the initial flat bed bathymetry of the schematized systems, the process of basin infilling was taken into account in all CC simulations forced with SLR by simply raising the initial lagoon bed levels by 50% of the SLR (by 2100).

Table 3.3. Adopted Climate Change forcing and sources.

CC Forcing	Adopted values	Source
SLR	+1m	IPCC AR5
Wave Height Variation (ΔH_s)	±8%	Hemer et al. (2013)
Wave Angle Variation ($\Delta\theta$)	±10°	Hemer et al.(2013)
Riverflow Variation (ΔR)	±40%	IPCC AR5

Two sets of CC simulations were undertaken for each system. The first to investigate how STI stability may be affected by CC driven variations in *individual system forcings*, and the second to investigate the impact of CC modified *key physical processes* (e.g. LST, *P*) on STI stability. In total, 15 CC simulations were undertaken for each of the 3 case study systems.

3.3   Results

3.3.1   Morphodynamic validation

3.3.1.1   Type 1 – Permanently open, locationally stable inlet (Case study: Negombo Lagoon)

Satellite images of Negombo lagoon consistently show a highly stable (locationally and cross-sectionally) inlet (Figure 3.8). Starting from the schematized flat bed bathymetry (Figure 3.9, left), the model correctly reproduces the locationally and cross-sectionally stable inlet behaviour seen in the satellite images. Some ebb and flood delta development is shown in the model result after 1 year. This is not unexpected as the initial flat bed bathymetry has to adjust to system forcing. The modelled annual longhsore sediment transport along the coastline is relatively small at 50,000 m³/yr and in agreement with reported values (Chandramohan et al., 1990).

Figure 3.8. Satellite images of Negombo lagoon, showing the locationally and cross-sectionally stable inlet behaviour (source: *Landsat*).

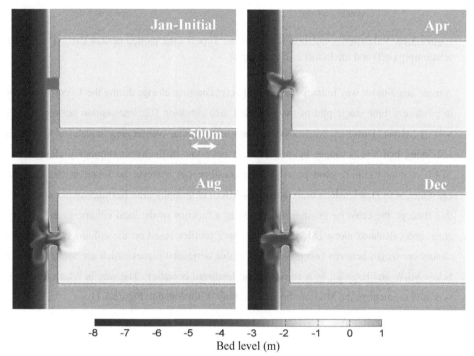

Figure 3.9. Validation model results showing the annual bed level evolution of the Type 1 inlet.

Model results also agree well with the empirical A-P equilibrium relationship (Figure 3.10, left) and the Escoffier curve (Figure 3.10, right). The modelled A and P values (per tidal cycle) lie well within the 95% confidence intervals and very close to the expected value line of the A-P relation for unjettied inlets presented by Jarrett (1976). Similarly, the modelled maximum inlet velocity and A values (per tidal cycle) lie around the stable root of the Escoffier curve generated for this case study. The Bruun criterion (r) value calculated using model derived P and M values is 233 (> 150), which also indicates a very stable inlet according to the Bruun stability criteria shown in Table 2.1.

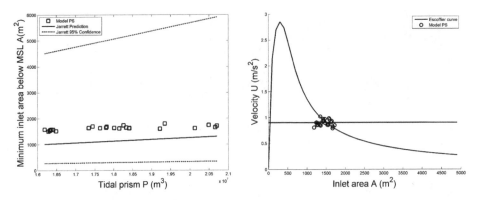

Figure 3.10. Validation simulation results of the Type 1 inlet plotted against the *A-P* equilibrium relationship (left) and the Escoffier curve (right).

A more quantitative way to track inlet and adjacent coastline change during the 1 year simulation is to produce a 'time stack' plot of the modelled zero elevation (i.e. interception between land and MWL) contour. However, while *Delft3D* (and other similar coastal area models) is very good at simulating bed level changes below MWL, it has some difficulty in simulating changes in the MWL contour itself. A good proxy for the coastline can however be found in the Momentary coastline (or in Dutch, Momentary Kustlijn - MKL) philosophy (van Koningsveld et al., 2004). In this concept, the coastline position is defined as a function of the sand volume in the near shore zone, and calculated along individual cross-shore profiles based on the volume of sand per unit alongshore length between two pre-defined, stable horizontal planes which are located above and below MLW and bounded by a fixed vertical landward boundary. The way in which this concept was used to calculate the MKL in the present study is illustrated in Figure 3.11.

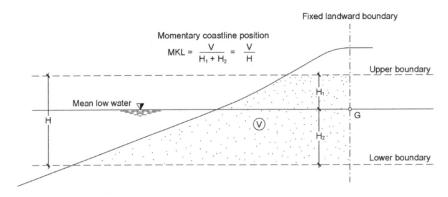

Figure 3.11. Schematic diagram showing the Momentary Coastline (MKL) concept.

The MKL position was calculated at the end of each month of the PS at every cross-shore grid line. The monthly MKL time stack plot thus obtained (Figure 3.12, left) shows that the model correctly reproduces a locationally and cross-sectionally stable inlet (after some initial widening of the inlet while the schematized bathymetry adjusts itself to the abruptly introduced forcing conditions). Figure 3.12 (right) shows the spatio-temporal evolution of the change in MKL relative to the initial MKL (i.e. Δ MKL), which re-confirms the modelled locationally and cross-sectionally stable inlet. Accretion (positive Δ MKL) is shown on both sides of the inlet, which is to be expected at this case study site because the wave directions alternate around the shore normal through the year.

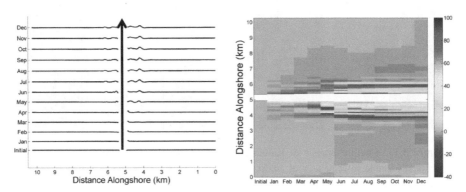

Figure 3.12. MKL (left) and Δ MKL (right) of the validation simulation for the Type 1 inlet.

3.3.1.2   Type 2 – Permanently open, alongshore migrating Inlet (Case study: Kalutara Lagoon)

Satellite images of the Kalutara lagoon show that inlet has historically migrated about 2 km southward in 3-4 years (annual migration of ~500m). Once the inlet reaches the southernmost point of the lagoon, a new, more hydraulically efficient inlet has traditionally been naturally or artificially created at the northern end of the lagoon, and the migration cycle then repeats (Figure 3.13). The PS for Kalutara inlet correctly reproduces the locationally unstable and cross-sectionally stable inlet behaviour seen in the satellite images (Figure 3.14). The modelled annual longshore sediment transport is 450,000 m^3 to the south while the migration rate is about 600m/yr to the south, both of which are in agreement with the reported values (GTZ, 1994; Perera, 1993).

The alongshore inlet migration distance (L) and speed (V_i) during the validation simulation are shown in Figure 3.15. Here, $L = 0$ denotes the original position of the inlet, and $L < 0$ indicates southward inlet migration. Inlet migration is high during the SW monsoon when higher waves are present and very small to non-existent during the rest of the year. The southward migration of the inlet is driven by higher southward LST during the SW monsoon (up to ~ 2700m^3/day). Although the deepwater waves are southerly during the SW monsoon, the combined effect of wave refraction over the large canyon near Kalutara and the coastline orientation of ~20^0 (anticlockwise from North) in the area results in a southward longshore transport.

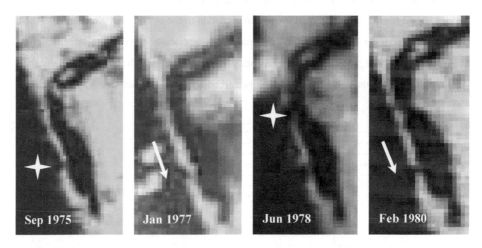

Figure 3.13. Satellite images of Kalutara lagoon, showing the locationally unstable and cross-sectionally stable inlet behaviour (source: *Landsat*). The asterisks indicate inlet location and the arrows indicate migration direction.

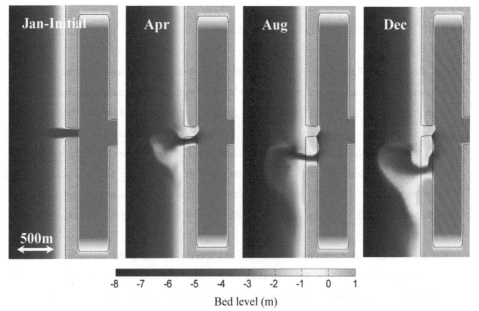

Figure 3.14. Validation model results showing the annual bed level evolution of the Type 2 inlet. The black line indicates the initial shoreline

As in the Type 1 inlet case above, model results are in good agreement with the *A-P* relationship (for unjettied inlets, Jarrett (1976) (Figure 3.16), consistent with the fact that, although this inlet is locationally unstable, it is still cross-sectionally stable, which is the only type of stability that the *A-P* relationship refers to. A traditional Escoffier type curve cannot be constructed for this case as system dimensions fall below its limits of applicability.

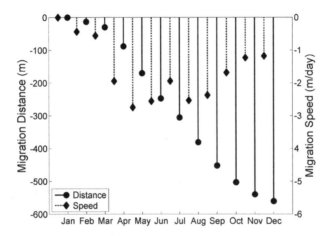

Figure 3.15. Modelled inlet migration distance and speed through the 1 year validation simulation of the Type 2 inlet.

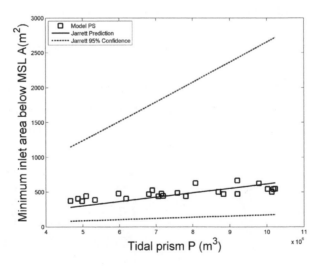

Figure 3.16. Validation simulation results of the Type 2 inlet plotted against the *A-P* equilibrium relationship.

The Bruun criterion (r) value calculated using model derived P and M values is 17.5 (< 20), which indicates an unstable inlet following Table 2.1. This result however implies that the 'unstable inlet' classification when $r < 20$ in the Bruun stability criteria applies more to locational stability rather than to cross-sectional stability.

The MKL plot (Figure 3.17, left) clearly shows the permanently open, southward migrating behaviour of this inlet. The rapid southward migration of the inlet during the SW monsoon and its relatively stationary behaviour during the rest of the year is illustrated by the Δ MKL plot (Figure 3.17, right).

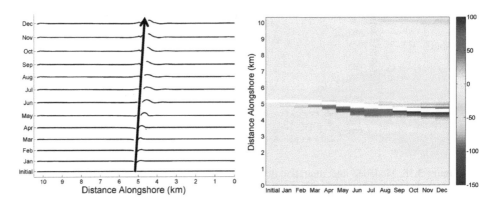

Figure 3.17. MKL (left) and Δ MKL (right) of the validation simulation for the Type 2 inlet.

3.3.1.3   Type 3 – Seasonally/Intermittently open inlet (Case study: Maha Oya River)

Satellite images of Maha Oya inlet show that the inlet intermittently closes whenever riverflow is small (Figure 3.18). Mostly, the closure occurs outside of the NE monsoon, which should be the case as when the Maha Oya catchment derives most of its annual runoff from the NE monsoon.

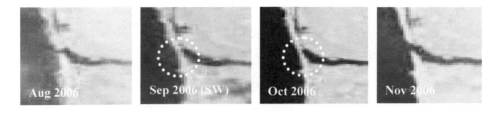

Figure 3.18. Satellite images of Maha Oya inlet, showing the locationally stable and cross-sectionally unstable inlet behaviour (source: *Landsat*).

The PS for Maha Oya inlet reproduces the locationally stable and cross-sectionally unstable inlet behaviour seen in the satellite images (Figure 3.19). The PS, which starts when riverflows are at its lowest, simulates inlet closure within one month. The modelled annual LST is 450,000 m³ to the North, which is in agreement with reported values (GTZ, 1994). The total closure of the inlet is further illustrated by the time evolution of the inlet cross-section and the inlet cross-sectional area shown in Figure 3.20 (left and right, respectively). Empirical relationships such as the *A-P* relationship and Escoffier curve are not valid for inlets with cross-section areas less than 500 m² (Byrne et al., 1980; Behrens et al., 2013) and were therefore not constructed for this case study.

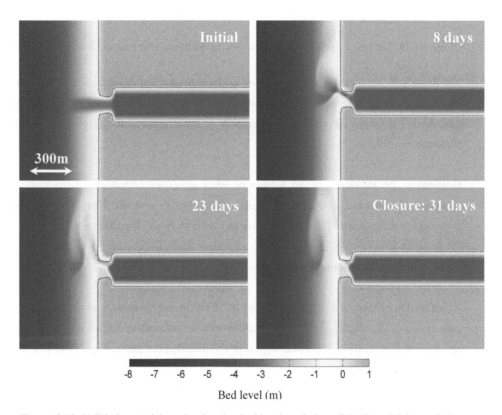

Bed level (m)

Figure 3.19. Validation model results showing bed level evolution of the Type 3 inlet until closure. The black line indicates the initial shoreline.

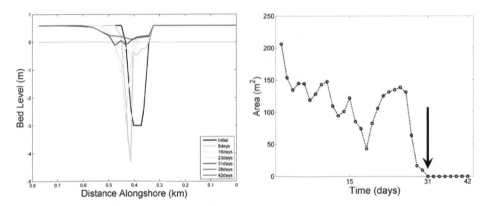

Figure 3.20. Time evolution of the inlet cross-section (left) and inlet cross-sectional area (right).

The Bruun criterion (r) value calculated using model derived P and M values is 2 (< 20), which also indicates an unstable inlet following Table 2.1. This result implies that an r value far lower than Bruun's threshold for unstable conditions ($r = 20$) is required for an inlet to be cross-sectionally unstable.

In summary, the PS results for all 3 types of inlets agree well with observed morphological behaviour and empirical relationships of inlet stability (where applicable), providing sufficient confidence in the respective PS models to move forward with the CC impact simulations.

3.3.2   Climate Change impact simulations

As outlined in Section 3.2.6, two separate sets of 15 CC impact simulations were undertaken for each system to investigate a) how STI stability may be affected by CC driven variations in individual system forcings, and b) the impact of CC modified key physical processes (e.g. LST, tidal prism) on STI stability.

3.3.2.1   Type 1 – Permanently open, locationally stable inlet (Case study: Negombo Lagoon)

CC driven variations in individual forcings and inlet stability

Table 3.4 summarises the CC driven variations (representing year 2100 conditions) applied to contemporary forcing in the 7 simulations (C1 to C7) undertaken for this analysis. The associated changes tidal prism (P) and annual longshore sediment transport volume (M) are also shown in 2nd last column of the table.

Table 3.4. Individual CC impact simulations of the Type 1 inlet: Forcing, associated changes in tidal prism P and annual longshore sediment transport M, and predicted future inlet type (ΔH_S: change in wave height; $\Delta\theta$: change in wave angle; ΔR: change in Riverflow).

CS	SLR 1m	ΔH_S +8%	ΔH_S -8%	$\Delta\theta$ +10°	$\Delta\theta$ -10°	ΔR +40%	ΔR -40%	Potential Change	Inlet Behaviour
C1	x							$M+,P+$	Type 1
C2		x						$M+$	Type 1
C3			x					$M-$	Type 1
C4				x				$M+$	Type 1
C5					x			$M+$	Type 1
C6						x		$P+$	Type 1
C7							x	$P-$	Type 1

The r values were calculated for all CC impact simulations, and are shown in comparison with that of the associated PS in Figure 3.21. Under all CC forcing conditions, r remains above 50, implying that the inlet will remain locationally and cross-sectionally stable in future. Thus, no individual CC forcing appears to be capable of changing the inlet Type of these inlets, which is reflected in the last column of Table 3.4. However, while the inlet remains stable, the 'level' of stability does change under some CC forcings. Particularly, in C4 and C5 when M increases due to changes in wave direction, r decreases from its PS value of 233 to below 150, implying a significant reduction in inlet stability from 'good' to 'fair' (C5) or 'fair to poor' (C4) according to Table 2.1. It is noteworthy that all other CC driven changes in individual forcings such as SLR, wave height and riverflow do not result in an appreciable change in the r value compared to the PS (i.e. r remains > 150, in Bruun's (1978) 'good' stability class).

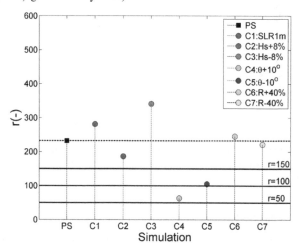

Figure 3.21. Bruun stability criterion for CC driven variations in individual forcings at the Type 1 inlet. The r value for the PS is also shown (left) for comparison.

CC driven variations in individual forcings and inlet-adjacent coastline

Apart from inlet stability, CC driven variations in system forcing may also affect the coastline adjacent to the inlet in two ways; changes in coastline position (e.g. mean coastline recession) and changes in spatial variability of the coastline. As discussed in Section 3.3.1.1, *Delft3D* computed coastline positions are not very reliable. Therefore, the Scale aggregated model for Inlet-interrupted Coasts (SMIC) presented by Ranasinghe et al. (2013) was used here to calculate potential coastline position changes under the CC forcings considered in C1 to C7. SMIC accounts for not only the Bruun effect of recession, but also the basin infilling effect, basin volume change effects and fluvial sediment supply change effects due to CC modified riverflow. However, SMIC doesn't provide any information on spatial variability of the coastline. In order to calculate this latter quantity, the MKL was calculated at the end of each simulation (including the PS). The (spatial) standard deviation of this end-MKL is representative of the alongshore variability of the coastline. These standard deviations are plotted around the SMIC predicted 'mean' coastline position in Figure 3.22 to illustrate variations in mean coastline position and spatial variability resulting from the considered CC driven variations in individual forcings.

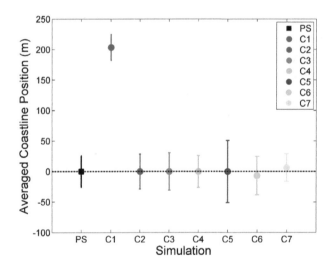

Figure 3.22. Changes in mean coastline position (positive = recession) and spatial coastline variability due to CC driven variations in individual forcings at the Type 1 inlet. The PS conditions are shown on the left for comparison.

Figure 3.22 shows that SLR is the major cause of mean coastline recession (up to ~200 m), which is not unexpected. Other CC driven changes in individual system forcing do not appear to result in significant coastline recession/progradation. The spatial variability of the coastline appears to differ significantly from the contemporary condition (i.e. PS) in the C5 simulation (spatial standard deviation of ~50 m, as opposed to ~25 m in the PS), when M increases due to a CC driven southerly rotation (of 10^0) in mean wave direction.

CC driven variations in physical processes and inlet stability

In this part of the analysis, strategic combinations of CC modified forcing were used to investigate the impact of CC modified key physical processes (e.g. longshore sediment transport, tidal prism, sea level rise) on the stability of the Type 1 inlet. The combinations of CC forcing adopted (in simulations C8-C15), associated changes in tidal prism (P), annual longshore sediment transport (M), and SLR are shown in Table 3.5. The r values calculated for this set of simulations are shown in Figure 3.23.

Table 3.5. Combined CC impact simulations of the Type 1 inlet: Forcing, associated changes in tidal prism P and annual longshore sediment transport M, and predicted future inlet type (ΔH_S: change in wave height; $\Delta\theta$: change in wave angle; ΔR: change in Riverflow).

CS	SLR 1m	ΔH_S +8%	ΔH_S -8%	$\Delta\theta$ +10^0	$\Delta\theta$ -10^0	ΔR +40%	ΔR -40%	Potential Change	Inlet Behaviour
C8		x		x		x		$M+,P+$	Type 1
C9		x		x			x	$M+,P-$	Type 1
C10	x	x		x		x		$M+,P+$	Type 1
C11	x	x		x			x	$M+,P-$	Type 1
C12			x		x	x		$M+,P+$	Type 1
C13			x		x		x	$M+,P-$	Type 1
C14	x		x		x	x		$M+,P+$	Type 1
C15	x		x		x		x	$M+,P-$	Type 1

Under all combined CC forcing conditions, r remains (at or) above 50, implying that the inlet will remain locationally and cross-sectionally stable in future under these forcing conditions. Thus, the inlet will remain as a Type 1 inlet regardless of CC driven variations in the governing physical processes, which is reflected in the last column of Table 3.5. However, the 'level' of stability decreases significantly (to around $r = 50$; from 'good' to 'fair to poor' in Table 2.1) when M increases (C8 and C9), and moderately (to around $r = 100$; from 'good' to 'fair') when the increase in M is accompanied by SLR (C10 and C11). When all other forcing stays the same, SLR generally shows a tendency to increase the stability level of Type 1 inlets (e.g. C10 *vs* C8; C11 *vs* C9; C14 *vs* C12; C15 *vs* C13). This is because with SLR the inlet channel becomes deeper, resulting in less tidal attenuation and thus an increased tidal prism. CC driven changes in riverflow do not appear to have any significant impact on the r value (e.g. C8 *vs* C9; C10 *vs* C11).

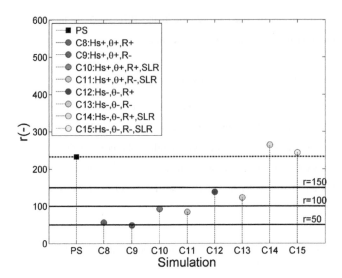

Figure 3.23. Bruun stability criterion for CC driven variations in physical processes at the Type 1 inlet. The *r* value for the PS is also shown (left) for comparison.

CC driven variations in physical processes and inlet-adjacent coastline

Similar to the individual forcings results, here too SLR shows the major impact on mean coastline recession (up to ~200 m), while CC driven variations in *M* and *P* do not result in much coastline recession (Figure 3.24).

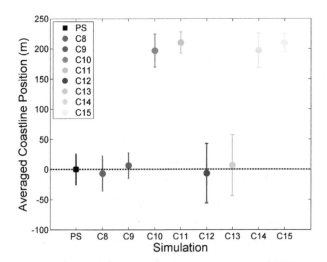

Figure 3.24. Changes in mean coastline position (positive = recession) and spatial coastline variability due to CC driven variations in physical processes at the Type 1 inlet. The PS conditions are shown on the left for comparison.

Similar to the individual forcing analysis, here too the highest coastline variability is indicated when M increases due to an anti-clockwise rotation of 10^0 in the wave direction (C12 and C13). Interestingly, the lower coastline variability in C14 and C15 relative to C12 and C13 (all simulations in which a CC driven 10^0 anti-clockwise rotation of waves is affected) indicates that SLR appears to have a damping effect of coastline variability.

In summary, locationally and cross-sectionally stable inlets will not change Type under CC, with r values always > 50. As such, Type 1 inlets appear to span the 3 stability classes 'fair to poor', 'fair, 'and 'good' in the Bruun stability criteria. The level of stability of Type 1 inlets could however decrease significantly (by up to two stability classes in the Bruun inlet stability criteria), particularly due to CC driven increases in annual longshore sediment transport. SLR, generally increases the level of inlet stability, and when acting in combination with CC driven increases in longshore transport, appears to compensate for the drops in inlet stability level by the latter. Mean coastline recession adjacent to the inlet can be as much as 200 m, largely due to the combination of the Bruun effect and the Basin infilling effect. Increased longshore sediment transport results in larger coastline variability relative to the present.

3.3.2.2   Type 2 – Permanently open, alongshore migrating Inlet (Case study: Kalutara Lagoon)

CC driven variations in individual forcings and inlet stability

The model predicted impacts of CC driven variation in individual forcings on the stability of the Type 2 inlet are shown in Table 3.6 and Figure 3.25, indicating that except in one case (C5), the inlet will not change Type.

Table 3.6. Individual CC impact simulations of the Type 2 inlet: Forcing, associated changes in tidal prism P and annual longshore sediment transport M, and predicted future inlet type (ΔH_S: change in wave height; $\Delta\theta$: change in wave angle; ΔR: change in Riverflow).

CS	SLR 1m	ΔH_S +8%	ΔH_S -8%	$\Delta\theta$ +10°	$\Delta\theta$ -10°	ΔR +40%	ΔR -40%	Potential Change	Inlet Behaviour
C1	x							$M+,P+$	Type 2
C2		x						$M+$	Type 2
C3			x					$M-$	Type 2
C4				x				$M+$	Type 2
C5					x			$M-$	**Type 1**
C6						x		$P+$	Type 2
C7							x	$P-$	Type 2

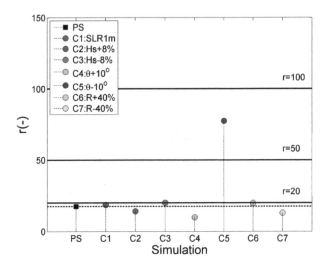

Figure 3.25. Bruun stability criterion for CC driven variations in individual forcings at the Type 2 inlet. The r value for the PS is also shown (left) for comparison.

The reduction of M due to a CC driven 10^0 anti-clockwise rotation of wave direction in C5 results in r increasing to > 50 (from its PS value of $17.5 < 20$) changing inlet Type to a locationally and cross-sectionally stable inlet (Type 1). The time evolution of modelled bed levels in this case are shown in Figure 3.26.

Increases/decreases in M due to changes in wave direction result in significant (~500 m/yr) increases/decreases in inlet migration distance (Figure 3.27; C4 and C5). CC driven variations in riverflow, wave height or SLR alone have insignificant impact on inlet stability (r varies in the range 5-20) and inlet migration distance (< 100 m/yr variation from present).

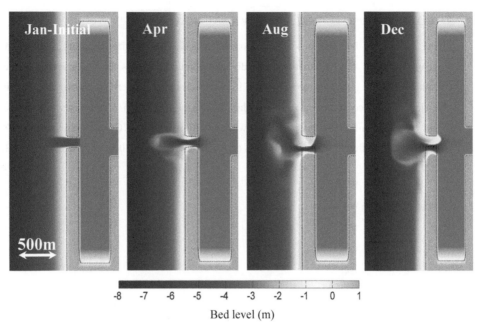

Figure 3.26. Modelled morphological evolution in Simulation C5 showing the Type 2 inlet turning into a locationally and cross-sectionally stable Type1 inlet when CC results in a 10^0 anti-clockwise rotation of wave direction. The black line indicates initial shoreline.

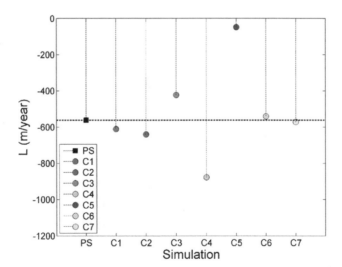

Figure 3.27. Inlet migration distance for CC driven variations in individual forcings at the Type 2 inlet. The migration distance for the PS is also shown (left) for comparison.

CC driven variations in individual forcings and inlet-adjacent coastline

Figure 3.28 shows the variations in mean coastline position and spatial variability resulting from the considered CC driven variations in individual forcings. SLR is the major driver of mean coastline recession (~120 m recession in C1), while changes in riverflow shows only marginal recession. Spatial coastline variability in this case is highest when M increases due to a 10^0 clockwise rotation of the wave direction or when riverflow increases (C4 and C6 respectively) but not significantly different from that of the PS (standard deviation of ~30 m relative to 20 m in PS).

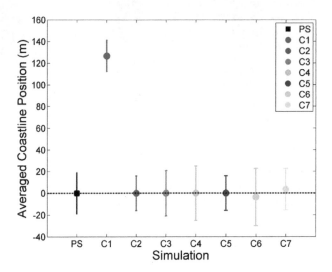

Figure 3.28. Changes in mean coastline position (positive = recession) and spatial coastline variability due to CC driven variations in individual forcings at the Type 2 inlet. The PS conditions are shown on the left for comparison.

CC driven variations in physical processes and inlet stability

The model predicted impacts of CC driven variations in physical processes on the stability of the Type 2 inlet are shown in Table 3.7 and Figure 3.29, indicating that when M decreases, regardless of SLR and riverflow effects, the inlet will change type to a locationally and cross-sectionally stable Type 1 inlet (C12-C15, where M decreases due to the 10^0 anti-clockwise rotation of wave direction and the 8% decrease in wave height). The r values in these 4 cases increase by up to 100 relative to the PS ($r = 17.5$), thus shifting the inlet by up to 3 stability classes in Bruun's stability criteria (i.e. from 'unstable' to 'fair to poor' or 'fair'). The time evolution of modelled bed levels in C12 and C15 are shown in Figure 3.30.

Table 3.7. Combined CC impact simulations of the Type 2 inlet: Forcing, associated changes in tidal prism P and annual longshore sediment transport M, and predicted future inlet type (ΔH_S: change in wave height; $\Delta\theta$: change in wave angle; ΔR: change in Riverflow).

CS	SLR 1m	ΔH_S +8%	ΔH_S -8%	$\Delta\theta$ +10°	$\Delta\theta$ -10°	ΔR +40%	ΔR -40%	Potential Change	Inlet Behaviour
C8		X		X		X		$M+,P+$	Type 2
C9		X		X			X	$M+,P-$	Type 2
C10	X	X		X		X		$M+,P+$	Type 2
C11	X	X		X			X	$M+,P-$	Type 2
C12			X		X	X		$M-,P+$	**Type 1**
C13			X		X		X	$M-,P-$	**Type 1**
C14	X		X		X	X		$M-,P+$	**Type 1**
C15	X		X		X		X	$M-,P-$	**Type 1**

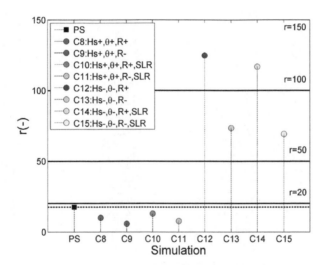

Figure 3.29. Bruun stability criterion for CC driven variations in physical processes at the Type 2 inlet. The r value for the PS is also shown (left) for comparison.

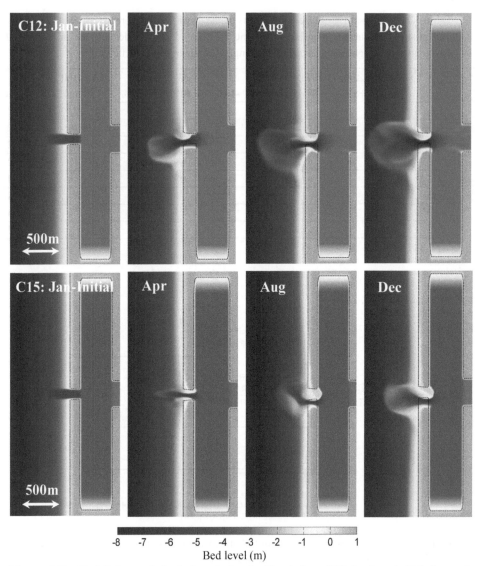

Figure 3.30. Modelled morphological evolution in Simulations C12 (top) and C15 (bottom) showing the Type 2 inlet turning into a locatinally and cross-sectionally stable Type 1 inlet when CC results in a decrease in annual longshore sediment transport. Black line shows initial shoreline.

Inlet migration distance increases significantly (~500 m/yr) in simulations C8-C11 where M increases due to the 10^0 clockwise rotation of wave direction and the 8% increase in wave height (Figure 3.31). Conversely, when M decreases in C12-C15 relative to the PS, annual inlet migration decreases by ~500 m, resulting in an insignificant migration rate ($< ~50$ m/yr).

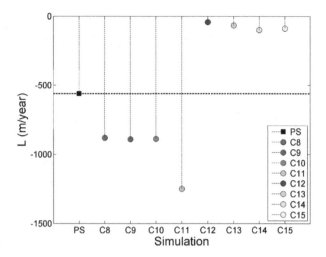

Figure 3.31. Inlet migration distance for CC driven variations in physical processes at the Type 2 inlet. The migration distance for the PS is also shown (left) for comparison.

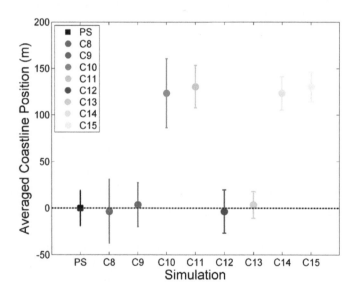

Figure 3.32. Changes in mean coastline position (positive = recession) and spatial coastline variability due to CC driven variations in physical processes at the Type 2 inlet. The PS conditions are shown on the left for comparison.

CC driven variations in physical processes and inlet-adjacent coastline

Figure 3.32 shows the variations in mean coastline position and spatial coastline variability computed for simulations C8-C15. Results show that SLR is the major driver of mean coastline recession (~120 m recession) as evidenced by the predictions for simulations C10, C11, C14 and

C15 relative to other simulations. Spatial coastline variability is significantly higher than that in the PS only in simulations C8 and C10 (standard deviations of ~35 m relative to ~20 m in the PS), when both M and P increase.

In summary, a reduction in annual longshore sediment transport, due mainly to CC driven variations in wave direction, and to a lesser degree in wave height, can increase the stability of an alongshore migrating Type 2 inlet so much that it changes into a locationally and cross-sectionally stable Type 1 inlet. CC driven increases/decreases in annual longshore sediment transport can increase/decrease inlet migration distance by ~100%. SLR does not appear to have any noticeable impact on the stability of Type 2 inlets. In all situations where the inlet remained a Type 2 inlet (with individual or combined CC forcing), r values stay in the range 5-20, implying a discrete sub-category (permanently open and alongshore migrating inlet) not explicitly mentioned in the Bruun stability criteria. SLR is the main driver of mean coastline recession adjacent to the inlet, which can be ~120 m, and is largely due to the Bruun effect. A significant increase in the spatial variability of the coastline relative to the present is predicted when both the tidal prism and annual longshore sediment transport increase concurrently.

3.3.2.3   Type 3 – Seasonally/Intermittently open, locationally stable inlets (Case study: Maha Oya river)

CC driven variations in individual forcings and inlet stability

The inlet stability analysis results with CC driven variations in individual forcings are shown in Table 3.8 and Figure 3.33, which indicate that inlet Type does not change under any of the tested forcing conditions. Simulation C4, where a 10^0 clockwise rotation of wave direction decreases M, shows a marginal increase of r to 8 (from the PS value of 2).

Table 3.8. Individual CC impact simulations of the Type 3 inlet: Forcing, associated changes in tidal prism P and annual longshore sediment transport M, and predicted future inlet type (ΔH_S: change in wave height; $\Delta\theta$: change in wave angle; ΔR: change in Riverflow).

CS	SLR 1m	ΔH_S +8%	ΔH_S -8%	$\Delta\theta$ +10^0	$\Delta\theta$ -10^0	ΔR +40%	ΔR -40%	Potential Change	Inlet Behaviour
C1	x							$M+,P+$	Type 3
C2		x						$M+$	Type 3
C3			x					$M-$	Type 3
C4				x				$M-$	Type 3
C5					x			$M+$	Type 3
C6						x		$P+$	Type 3
C7							x	$P-$	Type 3

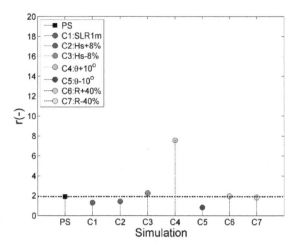

Figure 3.33. Bruun stability criterion for CC driven variations in individual forcings at the Type 3 inlet. The *r* value for the PS is also shown (left) for comparison.

The time till inlet closure for the various simulations is shown in Figure 3.34, indicating that the time till closure is more or less unchanged from its PS value except in C4 where the inlet remains open for 46 days (i.e. almost a 50% increase relative to the PS).

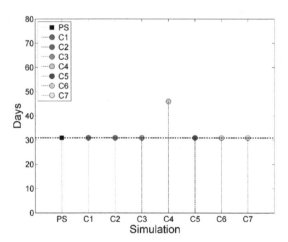

Figure 3.34. Time till inlet closure for CC driven variations in individual forcings at the Type 3 inlet. The time till closure for the PS is also shown (left) for comparison.

As these simulations were continued only until inlet closure (~one month) coastline recession and spatial variability computations, which are annual quantities, were not undertaken.

CC driven variations in physical processes and inlet stability

Inlet stability analysis of the simulations with CC driven variations in physical processes also show that the Type 3 inlet will remain as such in future (Table 3.9), with r never exceeding 10 (Figure 3.35). Thus, the inlet always remains in the 'unstable' class of Bruun's inlet stability criteria. However, when CC forcing results in a decreased M, the time till inlet closure increases by up to 200% (C12-C15) (Figure 3.36). The time till inlet closure is highest when P increases while M decreases (C12). Comparison of the results of C10 and C11 with those of C8 and C9, respectively, indicate that when CC results in an increased M, SLR could promote faster inlet closure (up to 50% faster relative to the PS).

Table 3.9. Combined CC impact simulations of the Type 3 inlet: Forcing, associated changes in tidal prism P and annual longshore sediment transport M, and predicted future inlet type (ΔH_S: change in wave height; $\Delta\theta$: change in wave angle; ΔR: change in Riverflow).

CS	SLR 1m	ΔH_S +8%	ΔH_S -8%	$\Delta\theta$ +10°	$\Delta\theta$ -10°	ΔR +40%	ΔR -40%	Potential Change	Inlet Behaviour
C8		X			X	X		$M+,P+$	Type 3
C9		X			X		X	$M+,P-$	Type 3
C10	X	X			X	X		$M+,P+$	Type 3
C11	X	X			X		X	$M+,P-$	Type 3
C12			X	X		X		$M-,P+$	Type 3
C13			X	X			X	$M-,P-$	Type 3
C14	X		X	X		X		$M-,P+$	Type 3
C15	X		X	X			X	$M-,P-$	Type 3

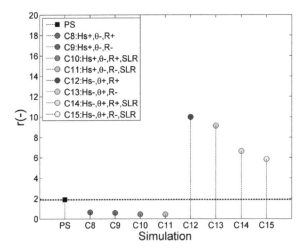

Figure 3.35. Bruun stability criterion for CC driven variations in physical processes at the Type 3 inlet. The r value for the PS is also shown (left) for comparison.

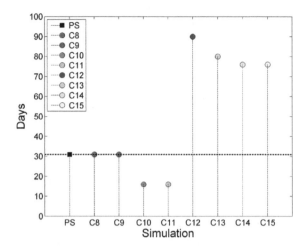

Figure 3.36. Time till inlet closure with CC driven variations in physical processes at the Type 3 inlet. The time till closure for the PS is also shown (left) for comparison.

In summary, seasonally/intermittently open inlets will not change Type due to CC driven variations in forcing/physical processes. The r value remained below 10 in all Type 3 inlet simulations, implying that for inlet closure, $r < 10$ is a necessary condition. Despite, not changing Type due to CC forcing, the time till inlet closure of these inlets may increase by up to 200% when CC results in a concurrent decrease in annual longshore sediment transport rate and an increase in tidal prism. In situations where the annual longshore sediment transport rate increases due to CC, SLR could result in faster inlet closure, decreasing the time till inlet closure by up to 50%.

3.4   Conclusions

A snap-shot simulation approach using a process based coastal area morphodynamic model has been applied to qualitatively assess CC impacts on the stability of Small Tidal Inlets (STIs). The modelling approach, which is intended specifically for data poor environments, was applied to three case study sites representing the main types of STIs: locationally and cross-sectionally stable inlets (Type 1, Negombo lagoon, Sri Lanka - permanently open, fixed in location); cross-sectionally stable, locationally unstable inlets (Type 2, Kalutara lagoon, Sri Lanka - permanently open, alongshore migrating); and locationally stable, cross-sectionally unstable inlets (Type 3, Maha Oya river, Sri Lanka - seasonally/intermittently open, fixed in location).

Schematized bathymetries that closely follow the key dimensions of the inlet systems were developed using available data, information and personal communications and used in all model simulations which were undertaken with *Delft3D*. Morphodynamic model validation simulations were undertaken using monthly averaged wave and riverflow time series constructed using available sparse data and reported values. In all 3 cases, successful model validation was achieved by comparing model results with available satellite images, and where appropriate, with empirical relationships such as the *A-P* relationship, Escoffier curve, and the Bruun inlet stability criteria.

The validated models were then used in two sets of simulations with CC modified forcing. The first set of simulations investigated the impact of CC driven variations in individual system forcings (i.e. SLR, wave height, wave direction, and riverflow) on STI stability, while the second set investigated the impact of CC driven variations in key physical processes (i.e. SLR, ebb tidal prism (including riverflow effects), and annual longshore sediment transport) on STI stability. Worst case estimates of SLR and CC modified wave/riverflow conditions were obtained from freely available, coarse resolution published data and used to force the CC impact model simulations. The main findings are listed below:

- Type 1 and Type 3 models will not change Type even under the most extreme CC forcing considered here. Type 2 inlets may change into Type 1 when CC results in a reduction in annual longshore sediment transport, due mainly to CC driven variations in wave direction, and to a lesser degree in wave height.

- Apart from Type changes, CC driven variations in system forcing and physical processes will affect the level of inlet stability and some behavioural characteristics. In general, CC driven variations in annual longshore sediment transport rates appear to be more relevant for future changes in inlet behaviour, rather than SLR as commonly believed.

- In Type 1 inlets, the Bruun Stability criterion *r* always remained greater than 50 (i.e. stable). The level of stability if Type 1 inlets could decrease significantly (by up to two stability classes in the Bruun inlet stability criterion; from 'good' to 'fair to poor'), due to CC driven increases in annual longshore sediment transport rate. Mean coastline recession adjacent to the inlet can be as much as 200 m, while CC driven increases in longshore sediment transport results in coastline variability that may be twice as much compared to the present.

- In Type 2 inlets, CC driven increases/decreases in annual longshore sediment transport volume can increase/decrease inlet migration distance by up to 100%. The *r* value for all Type 2 inlets (excepting when the inlet changed to Type 1) remained between 5 and 20 (i.e. unstable). SLR results in a mean coastline recession adjacent to the inlet of ~120 m, and a near doubling of the spatial variability of the coastline relative to the present is predicted when both the ebb tidal prism and annual longshore sediment transport volume increase concurrently.

- In Type 3 inlets, the r value remained below 10 (i.e. unstable) under all situations simulated, implying that $r < 10$ is a necessary condition for inlet closure. The time till closure of these inlets may increase by up to 200% when CC results in a concurrent decrease in annual longshore sediment transport rate and an increase in ebb tidal prism. When annual longshore sediment transport rate increases due to CC, SLR could decrease the time till inlet closure by up to 50% compared to present conditions.

- Based on the results of this study, inlet Types can be linked with the Bruun stability criteria to develop a more descriptive inlet classification scheme as shown in Table 3.10.

Table 3.10. Classification scheme for inlet Type and stability conditions.

Inlet Type	$r = P/M$	Bruun Classification
Type 1	> 150	Good
	100 - 150	Fair
	50 - 100	Fair to Poor
	20 - 50	Poor
Type 2	10 - 20	Unstable (open and migrating)
Type 2/3	5 - 10	Unstable (migrating or intermittently closing)
Type 3	0 - 5	Unstable (intermittently closing)

References

Behrens, D.K., Bombardelli, F.A., Largier, J.L., Twohy, E., 2013. Episodic closure of the tidal inlet at the mouth of the Russian River - A small bar-built estuary in California. Geomorphology, http://dx.doi.org/10.1016/j.geomorph.2013.01.017.

Bruun, P., 1978. Stability of tidal inlets - theory and engineering. Developments in Geotechnical Engineering. Elsevier Scientific, Amsterdam, 510p.

Byrne, R., Gammisch, R., Thomas, G., 1980. Tidal prism-inlet area relations for small tidal inlets. Proceedings of the 21st International Conference on Coastal Engineering, ASCE, New York, pp. 2517-2533.

Chandramohan, P., Nayak, B.U., 1990. Longshore - transport model for South Indian and Sri Lankan coasts. Journal of Waterway, Port, Coastal, and Ocean Engineering, 116, 408-424.

GTZ., 1994. Longhsore sediment transport study for the South West coast of Sri Lanka. Project Report. 25p.

Hemer, M., Fan, Y., Mori, N., Semedo, A., Wang, X.L., 2013. Projected changes in wave climate from a multi-model ensemble. Nature Climate Change, 3, 471-476.

IPCC, 2013. Summary for Policymakers. In: Climate Change 2013: The Physical Science Basis. Contribution of Working Group I to the Fifth Assessment Report of the Intergovernmental Panel on Climate Change [Stocker, T.F., D. Qin, G.-K. Plattner, M. Tignor, S.K. Allen, J. Boschung, A. Nauels, Y. Xia, V. Bex and P.M. Midgley (eds.)]. Cambridge University Press, Cambridge, United Kingdom and New York, NY, USA.

Jarrett, J.T., 1976. Tidal prism – inlet area relationships. Technical Report GITI No.3, CERC, U.S. Army Engineer Waterways Experiment Station, Vicksburg, MS.

Lesser, G.R., Roelvink, J.A., van Kester, J.A.T.M., Stelling, G.S., 2004. Development and validation of a three-dimensional morphological model. Coastal Engineering, 51 (8-9), 883-915.

O'Brien, M.P., 1931. Estuary and tidal prisms related to entrance areas. Civil Engineering, 1(8), 738-739.

Perera, J.A.S.C., 1993. Stabilization of the Kaluganga river mouth in Sri Lanka. M.Sc Thesis Report. International Institute for Infrastructural Hydraulic and Environmental Engineering, Delft, The Netherlands, 97p.

Ranasinghe, R., Swinkels, C., Luijendijk, A., Roelvink, D., Bosboom, J., Stive, M., Walstra, D., 2011. Morphodynamic upscaling with the MORFAC approach: Dependencies and sensitivities. Coastal Engineering, 58, 806-811.

Ranasinghe, R., Duong, T.M., Uhlenbrook, S., Roelvink, D., Stive, M., 2013. Climate change impact assessment for inlet-interrupted coastlines. Nature Climate Change, 3, 83-87, DOI.10.1038/NCLIMATE1664.

Roelvink, J.A., 2006. Coastal morphodynamic evolution techniques. Coastal Engineering, 53, 277-287.

University of Moratuwa., 2003. Engineering study on the feasibility of dredging the Negombo Lagoon to improve water quality. Final Report. Part II: Technical & Enviromental Feasibility.

van Koningsveld, M., Mulder, J.P.M., 2004. Sustainable coastal policy developments in the Netherlands. A systematic approach revealed. Journal of Coastal Research, 20(2), 375-385. West Palm Beach (Florida), ISSN 0749-0208.

van Rijn, L.C., 1993. Principles of sediment transport in rivers, estuaries and coastal seas. Part 1. AQUA Publications, NL. 700p.

Wijeratne, E.M.S., 2002. Sea level measurements and coastal ocean modelling in Sri Lanka. Proceedings of the 1st scientific session of the National Aquatic Resources Research and Development Agency, Sri Lanka. 18p.

Zubair, L., Chandimala, J., 2006. Epochal changes in ENSO – streamflow relationships in Sri Lanka. Journal of Hydrometeorology, 7(6), 1237-1246.

CHAPTER 4

ASSESSING CLIMATE CHANGE IMPACTS ON THE STABILITY OF SMALL TIDAL INLETS IN DATA RICH ENVIRONMENTS

4.1   Introduction

This Chapter demonstrates the application of the process based snap-shot modelling approach for *data rich environments* proposed in Section 2.4.1. The approach is applied to the 3 case study sites representing the 3 main STI types described Section 3.2.1.

Following Ruessink and Ranasinghe's (2014) ensemble modelling approach described in Section 2.4.1 and Figure 2.10 therein, a modified framework shown in Figure 4.1 was adopted in this study. The main deviations from the comprehensive framework suggested by Ruessink and Ranasinghe (2014) are: (1) only one GHG scenario (the high end A2 scenario) was considered, and (2) only one RCM was used. Thus, the GHG and RCM uncertainties are not accounted for in this study. This is mainly due to budgetary constraints of the study and therefore was unavoidable.

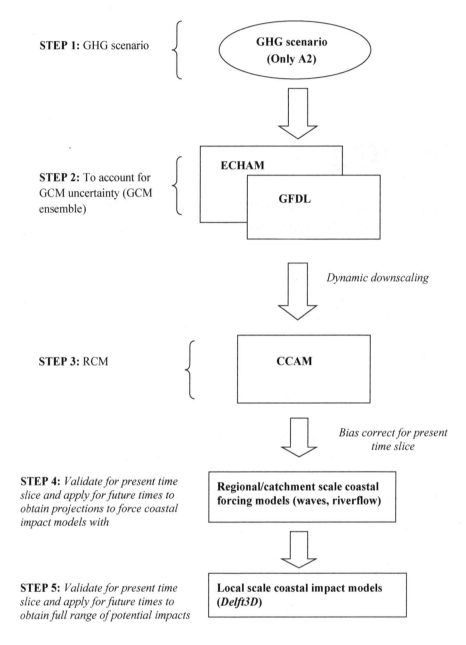

Figure 4.1. Modelling framework adopted in the present study following the comprehensive ensemble modelling framework proposed by Ruessink and Ranasinghe (2014).

4.2   Dynamic downscaling

All downscaled climate variables were derived from the stretched grid model CCAM (Conformal Cubic Atmospheric Model) which is owned and operated by CSIRO, Australia. CCAM is a semi-implicit, semi-Lagrangian atmospheric climate model based on a conformal cubic grid (McGregor and Dix, 2008). Although CCAM is a global atmospheric model, it allows a variable resolution grid which enables a finer grid resolution over the target area at the expense of a coarser resolution on the opposite side of the globe. In this way, CCAM can be used for regional climate experiments without imposing lateral boundary conditions. The variable resolution grid used to derive the downscaled climate variables over Sri Lanka for this study is shown in Figure 4.2. In this application CCAM employed 18 vertical levels (ranging from 40 m to 35 km. The grid used in the CCAM application for this study resulted in a resolution of about 65 km over Sri Lanka. The model was forced with Sea Surface Temperatures taken from two of the CMIP3 General Circulation Models (ECHAM and GFDL) which performed well in the target area. CCAM output including winds, surface temperature, atmospheric pressure, radiation, ocean temperature etc. was thus obtained for the 1981-2000 (present) and 2081-2100 time slices at a temporal resolution of 6 hours.

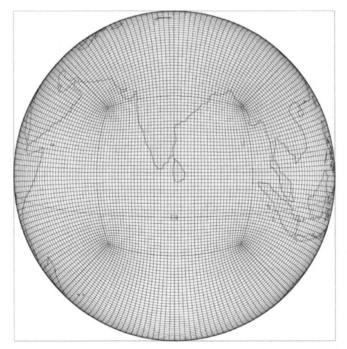

Figure 4.2. The variable resolution conformal cubic grid in the CCAM simulations.

4.3   Regional/catchment scale coastal forcing models

4.3.1   Riverflow

The CCAM output over Sri Lanka was used in a hydrologic model to derive riverflow estimates for the present (1981-2000) and future (2081-2100) (Mahanama and Zubair, 2011). The 6-hourly surface meteorological forcings used included shortwave radiation, longwave radiation, total precipitation, convective precipitation, surface pressure, air temperature, specific humidity, and wind for the two different periods. The hydrologic model used was the Catchment Land Surface Model (CLSM: Koster et al., 2000; Ducharne et al., 2000). CLSM is a macroscale hydrologic model that balances both surface water and energy at the Earth's land surface. One distinguishing characteristic of CLSM is that it considers irregularly shaped, topographically delineated, hydrologic catchments as the fundamental element on the land surface for computing land surface processes. The CLSM has already been successfully implemented in Sri Lanka (Mahanama et al., 2008) using bias corrected reanalysis meteorological forcings. For this study, the CLSM was forced in offline mode using the CCAM downscaled surface meteorological forcings to generate riverflows into the 3 case study lagoons.

Available gridded precipitation data were used for bias correcting the downscaled ECHAM and GFDL precipitation hindcasts for the present time slice, which were then used in CLSM to simulate riverflows. Observed monthly riverflows from 22 gauge stations across Sri Lanka for the period 1979-1993 were used for validating CLSM for the hindcast period 1981-2000. As the ECHAM and GFDL projections for the 3 case study lagoons were very similar, only GFDL projections were used to construct the annual cycle of riverflows to use as future forcing in the Coastal impact model, *Delft3D*. Here *Delft3D* was used with a MORFAC of 13 to ensure the representation of the spring-neap cycle in the CC impact assessments (see Section 4.4.1 below), and therefore, 13-day averaged riverflows were used to construct the average annual riverflows (Figure 4.3) to force the process based snap-shot model simulations described below in Section 4.4. In general, by 2100 (relative to the present), riverflow is projected to decrease by about 41% and 32% at Negombo lagoon and Kalutara lagoon respectively, while an increase of about 72% is indicated for Maha Oya river.

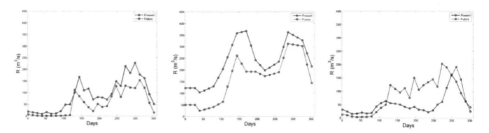

Figure 4.3. Contemporary and year 2100 riverflow time series for Negombo lagoon (left), Kalutara lagoon (middle), and Maha Oya river (right).

4.3.2   Waves

CCAM winds were used to force two nested spectral wave models for 1981-2000 (hindcast) and 2081-2100 (future) time slices (Bamunawala, 2013). Due to the similarity between CCAM downscaled ECHAM and GFDL winds in the study area, only CCAM-GFDL winds were used in this analysis. For the generation of far field waves, WAVEWATCH III (Tolman, 2009) was used (Latitudes N22^0-S7^0; Longitudes E65^0-E95^0). SWAN (Booij et al., 1999) was used in the near field from about 50 m depth to the coastline extending from Galle to Puttalam along the SW coast (see Figure 3.1 for locations). Modelled wave conditions for the hindcast period were compared against available deep water wave data off Colombo. The bias correction required to ensure a good model/data comparison was then determined and applied to the future projected wave conditions with the commonly adopted assumption that present-day biases between model and reality will remain the same in future (Charles et al., 2012; Wang et al., 2014). SWAN model output was then extracted at 20 m depth offshore of the 3 case study sites to use as boundary forcing in the process based snap-shot model simulations described in Section 4.4 below. As the process based model *Delft3D* was used with a MORFAC of 13 to ensure the representation of the spring-neap cycle in the CC impact assessments (see Section 4.4.1 below), 13-day averaged wave heights and directions were used to construct the average annual wave conditions for model forcing (Figure 4.4).

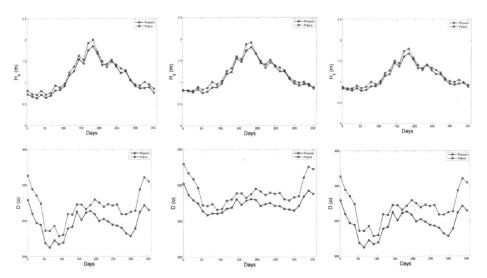

Figure 4.4. Contemporary and year 2100 wave time series (at 20m depth) for Negombo lagoon (left), Kalutara lagoon (middle), and Maha Oya river (right). Significant wave height (H_s) (top), mean wave direction ($D = \theta$) (bottom).

4.4   Coastal Impact modelling

The process based coastal area model *Delft3D* was used for all simulations undertaken in this study. The model is fully described by Lesser et al., (2004) and is therefore not described here. For the 3 case study applications of the model in this study, identical wave and flow domains which were large enough to avoid any boundary problems affecting the area of interest were created (Figure 4.5). High resolution (~10x10 m) grid cells were used in the (approximate) surf zone and inlet channel for all 3 study areas to ensure that key physical processes in the vicinity of the inlet entrance and channel were accurately resolved by the model. Good measured bathymetries were available for all 3 case study sites.

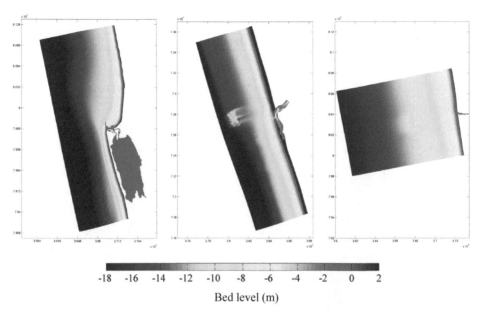

Bed level (m)

Figure 4.5. Wave/flow grids for Negombo (left), Kalutara (middle) and MahaOya (right)

4.4.1   Model validation

Hydrodynamic validation

First the models were validated against measured water level and velocity data inside the lagoons. Water level and velocity measurements for Negombo lagoon were available from a previous study. Two pressure sensors were deployed in Kalutara lagoon and Maha Oya river to collect water level data for this study specifically. Unfortunately however, due to problems with data acquisition, water level data at Kalutara lagoon was only captured intermittently, while the sensor deployed at Maha Oya river was lost. Therefore, hydrodynamic model validation could only be undertaken for Negombo and Kalutara lagoons. The hydrodynamic validation simulations were undertaken with only tidal and riverflow forcing as wave effects are minimal within the 3 case study STIs. Tidal forcing constituted of astronomical tides composed of the 6 main tidal constituents in the area (M2, S2, N2, K2, K1, O1), and riverflow was introduced as a time series based on available measurements. Morphological updating was turned off in these short-term simulations. The validation periods and data are shown in Table 4.1. The measurement locations are shown in Figure 4.6. A Chezy friction coefficient of $65m^{1/2}/s$, eddy viscosity of $1m^2/s$ and hydrodynamic time step of 6 seconds were adopted in all 3 hydrodynamic validation simulations.

Table 4.1. Data used for hydrodynamic model validation at the case study sites.

STI System	Data type	Data period
Negombo lagoon	Water level	01 - 30 Oct 2002
	Velocity	02 - 03 Oct 2002
Kalutara lagoon	Water level	13 - 26 Feb 2013

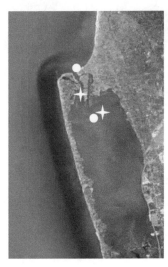

Figure 4.6. Measurement locations of model validation data in Negombo lagoon (left) and Kalutara lagoon (right). Filled white circles: water level observation points (Negombo - S1: ocean side, S3: inside lagoon; Kalutara - K1: inside lagoon). Asterisks: velocity observation points (Negombo - CM1: ocean side, CM2: inside lagoon).

Table 4.2. Model/data comparison statistics for the hydrodynamic validation simulations.

Negombo lagoon					
Water level	S1		S3		
	RMSE	R²	RMSE	R²	
	0.0325	0.9747	0.0312	0.8355	
Current	CM1		CM2		
Current velocity	0.1027	0.7319	0.0598	0.4397	
Current direction	10.17	0.6890	14.59	0.5628	
Kalutara lagoon					
	Feb 15		Feb 20		Feb 23
Water level	K1		K1		K1
	RMSE	R²	RMSE	R²	RMSE	R²
	0.1155	0.8668	0.0776	0.9872	0.0538	0.8747

The model/data comparisons for Negombo and Kalurara lagoons (Figures 4.7 and 4.8, and Table 4.2) are reasonably good, providing sufficient confidence in the models to proceed with the morphodynamic simulations.

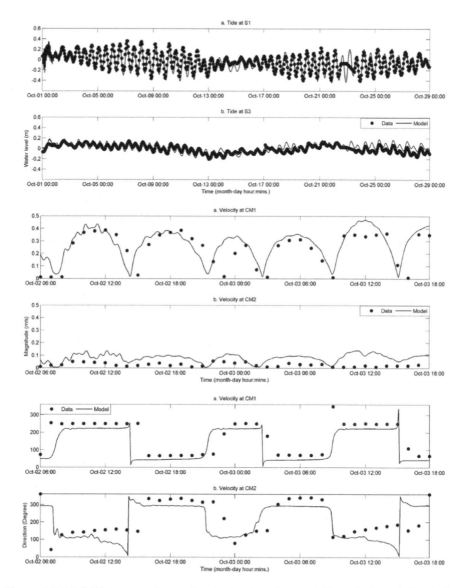

Figure 4.7. Model/data comparisons of water levels and currents (magnitude and direction) at Negombo lagoon.

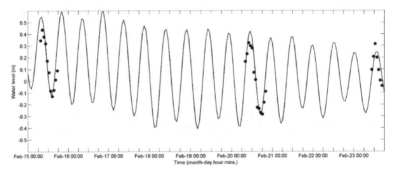

Figure 4.8. Model/data comparison of water levels at Kalutara lagoon

Morphodynamic validation

Morphodynamic validation was achieved in much the same way as it was done with the data poor approach adopted in Chapter 3. For Negombo lagoon and Kalutara lagoon, one year morphodynamic simulations were undertaken, while for Maha Oya river the morphodynamic simulation was continued only until the inlet closed. Models were forced with astronomical tides for which constituents were obtained from Wijeratne (2002). All model parameter values, except the MORFAC value and the Hydrodynamic spin up time, were the same as those indicated in Table 3.2. A MORFAC of 13 was used in these simulations in order to capture two spring-neap cycles (29 days) of hydrodynamic forcing within a 1 year morphodynamic simulation. A hydrodynamic spin up time of 24 hrs was used in these simulations to ensure that model velocities were stable before sediment transport and morphological computations were undertaken. Modelled bed level changes and satellite images for the 3 systems are shown in Figures 4.9-4.11. Modelled and measured (reported) annual LST (or M) in the vicinity of the 3 inlets are shown in Table 4.3. M was calculated following the approach described in Section 3.2.5.

Table 4.3. Modelled and measured (reported) annual LST (M) in the vicinity of the 3 case study inlets, the modelled Bruun criterion r and corresponding Bruun stability classification and inlet Type following Table 3.10.

STI system	Reported M (m³/yr)	Modelled M (m³/yr)	$r =$ P/M	Bruun stability classification	Inlet Type
Negombo Lagoon	20,000 S	42,000 S	221	Good	1
Kalutara Lagoon	500,000 S	562,000 S	11	Unstable	2
Maha Oya River	500,000 N	450,000 N	1	Unstable	3

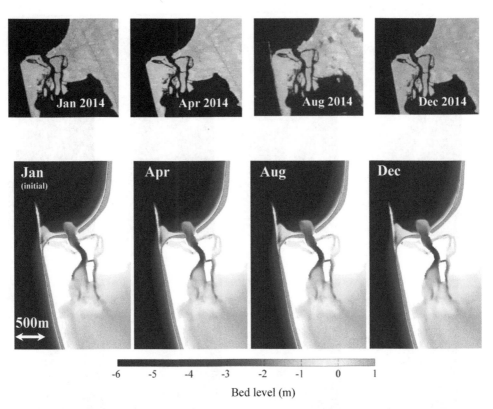

Figure 4.9. Satellite images (top (source: *Landsat*)) and validation model results (bottom) of the annual bed level evolution of Negombo lagoon, showing the observed and modelled locationally and cross-sectionally stable inlet behaviour. The black line in the model results indicates the initial shoreline position.

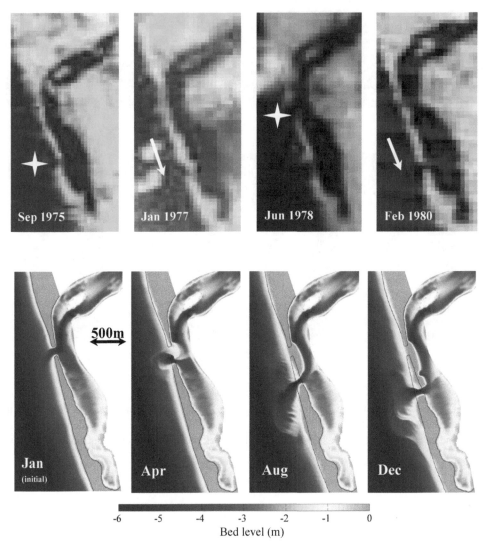

Figure 4.10a. Satellite images (top (source: *Landsat*)) and validation model results (bottom) of the annual bed level evolution of the Kalutara lagoon inlet, showing the observed and modelled locationally unstable and cross-sectionally stable inlet behaviour with the model correctly reproducing the observed southward migration of about 500 m/yr (see also Figure 4.10b). The black line in the model results indicates the initial shoreline position.

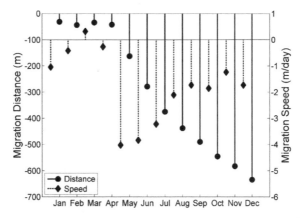

Figure 4.10b. Modelled inlet migration distance and speed through the 1 year validation simulation of Kalutara lagoon inlet showing that the model correctly reproduces observed southward (negative = southward) migration rate of ~500 m/yr and higher migration speeds during the SW monsoon.

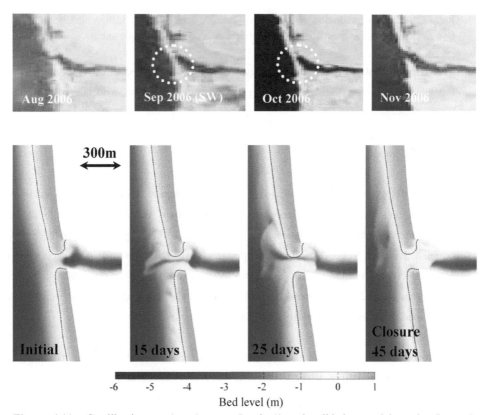

Figure 4.11a. Satellite images (top (source: *Landsat*)) and validation model results (bottom) showing the bed level evolution of the Maha Oya river inlet, showing the observed and modelled locationally stable and cross-sectionally unstable inlet behaviour. The white dashed circles in the satellite images indicate inlet closed situations. Black line in model results show initial coastline.

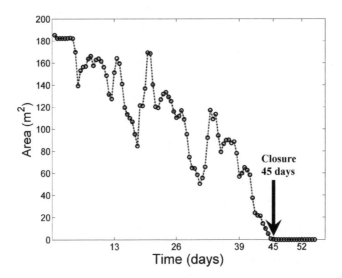

Figure 4.11b. Time evolution of inlet cross-sectional area in the validation simulation of Maha Oya river showing that the model correctly reproduces observed inlet closure.

In summary, the above results show that the model is able to reproduce contemporary observed behaviour of the 3 case study STIs providing sufficient confidence in the model to proceed with CC impact assessments.

4.4.2   CC impact assessment

For each STI system, the validated model was then implemented via snap-shot simulations to investigate future CC impacts on the system. These simulations were also undertaken for the same duration as the validation simulations, or, in the case of Maha Oya river, until inlet closure occurred. For each STI, CC modified riverflow and wave forcing were implemented using the projected forcing shown in Figures 4.3 and 4.4. A worst case SLR of 1m (by 2100 relative to the present) was applied at all 3 systems. The tidal forcing of all CC impact simulations were the same as that used in the respective validation simulations.

Due to the spatially non-uniform bathymetries of the systems, SLR driven basin infilling was implemented differently compared to the simple spatially uniform raising of the lagoon/inlet bed level method used in the flat-bed schematized models described in Chapter 3. Here, the bed levels of the initial measured bathymetry were changed to accommodate the basin infill volume (calculated in exactly the same way as described in Section 2.4.1, i.e. total infill volume = 0.5 x SLR x A_b, where A_b = surface area of lagoon, or basin) such that the shape of the present and future basin hyspometry curves were more or less the same. Basin hypsometry is the relationship between the basin depth (h_b) (measured from surface to the bottom, elevation = 0 at surface) and the basin area (A_b) (also measured from bottom to surface, with area = 0 at the bottom) (Boon and Byrne, 1981). To estimate the bed level changes required to represent basin infilling, first it is assumed that at all grid points:

$$h_{b,f} = (h_{b,p} + SLR) - \Delta h \tag{4.1}$$

where: Δh is assumed to follow the general depth transfer function given by,

$$\Delta h = a'(h_{b,p} + SLR) \tag{4.2}$$

where: a' is a coefficient, of which the optimal value is found via iteration. Subscripts 'p' and 'f' represent *present* and *future* respectively.

As an example, the year 2100 basin hypsometry curve calculated for Kalutara lagoon using the above approach is shown in Figure 4.12, together with the contemporary hypsometry curve.

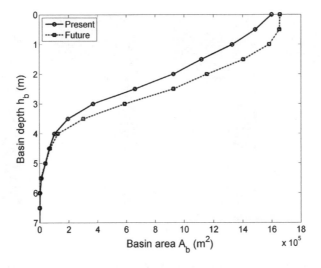

Figure 4.12. Present and future (year 2100) basin hypsometry curves indicative of the implemented bed level changes in Kalutara lagoon to represent SLR driven basin infilling.

Negombo Lagoon

The modelled future morphological changes over one year for the Type 1 Negombo lagoon are shown in Figure 4.13. For easy comparison, the validation simulation results for this STI shown in Figure 4.9 are also reproduced in Figure 4.13 (top panels). Model results show that this STI will remain a locationally and cross-sectionally stable inlet by 2100. The r value however decreases to 75, from its present value of 221. According to Bruun's inlet stability classification, this implies that the level of stability of the inlet will decrease from 'good' to 'fair to poor'.

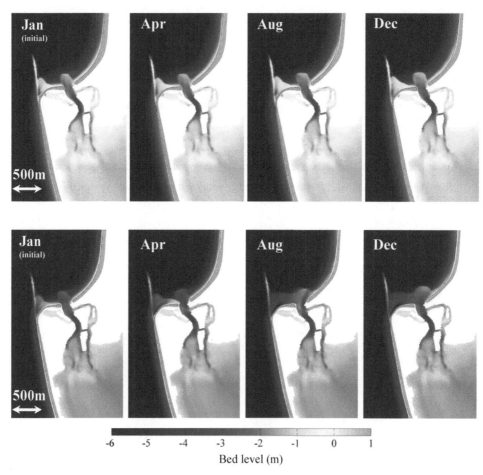

Figure 4.13. Modelled morphological changes for Negombo lagoon over a one year period; under contemporary forcing conditions (top) and CC modified year 2100 forcing conditions (bottom).

Kalutara lagoon

The modelled future morphological changes over one year for the Type 2 Kalutara lagoon are shown in Figure 4.14, together with corresponding validation simulation results. Model results show that Kalutara lagoon will remain a permanently open, alongshore migrating Type 2 STI by year 2100. However the migration distance doubles to ~1200 m, while the r value decreases to 6 from its present value of 11. These changes can be directly attributed the future increase in southward M due the CC driven clockwise rotation of waves (see Figure 4.4).

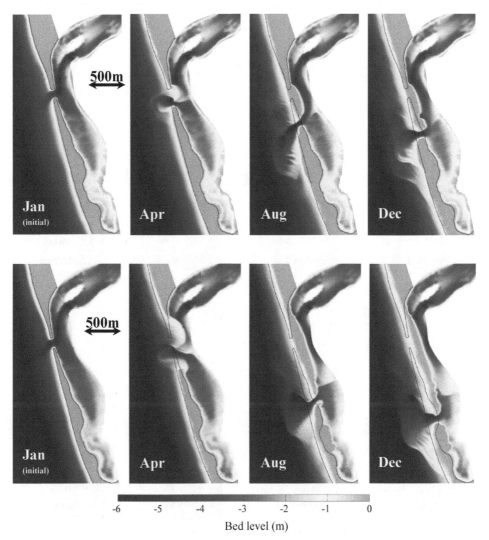

Figure 4.14. Modelled morphological changes for Kalutara lagoon over a one year period; under contemporary forcing conditions (top) and CC modified year 2100 forcing conditions (bottom).

Maha Oya river

The modelled future morphological changes for the Type 3 Maha Oya river are shown in Figure 4.15, together with corresponding validation simulation results. Model results show that this

STI will remain an intermittently open, locationally stable Type 3 STI by year 2100. However the time until inlet closure increases by more than 50% from its modelled present value of 45 days to 78 days, while the *r* value slightly increases to 5 from its present value of 1. These changes in system behaviour is due to the combined effect of the future increase in annual riverflow (see Figure 4.3) and the smaller northward *M* resulting from the CC driven clockwise rotation of waves (see Figure 4.4).

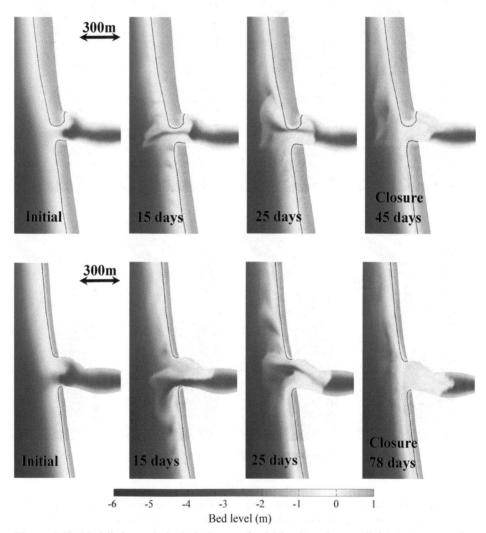

Figure 4.15. Modelled morphological changes for Maha Oya river until inlet closure; under contemporary forcing conditions (top) and CC modified year 2100 forcing conditions (bottom).

4.4.3 Relative contributions of CC driven variations in system forcing to inlet stability

For each case study, four additional simulations where CC forcing was sequentially removed (Simulations R2-R5; R1 being the above discussed 'all inclusive' CC impact simulation) were undertaken to investigate the relative contribution of the different CC forcings to future inlet stability. The CC forcings implemented in each simulation are shown in Table 4.4. Note that when CC modified future forcing is not used in a certain simulation (e.g. future riverflow R_f in R2), the present riverflow is still used for the year 2100 snap-shot simulation (i.e. representing a scenario where there is no change in riverflow due to CC). Also, basin infilling was not included in simulations that excluded SLR (i.e. R5). When SLR is implemented (Simulations R1-R4), it was specified as 1m. This set of simulations can be used to determine the effect of CC driven changes in each of the system forcings on future STI behaviour. For example, the difference between R2 and R1 would be indicative of the isolated effect CC driven changes in annual riverflow would have on STI behaviour, while differences between R5 and R1 would provide insights on the effect of SLR.

Table 4.4. Forcing conditions implemented in the different CC forcing simulations. Subscript 'f' indicates *future* conditions.

	SLR	H_{Sf},θ_f	R_f
R1	x	x	x
R2	x	x	
R3	x		x
R4	x		
R5		x	x

The main results from this set of simulations are summarised in Table 4.5. The results of the validation simulation (R1) are also shown for easy comparison.

Table 4.5. Model predicted year 2100 STI types and inlet behavioural characteristics in response to different CC forcings.

Simulation	Negombo lagoon		Kalutara lagoon			Maha Oya river		
	r	Type	r	Migration distance(m)	Type	r	Time till closure(days)	Type
R1	75	Type 1	6	1210	Type 2	5	78	Type 3
R2	82	Type 1	7	1183	Type 2	4	72	Type 3
R3	128	Type 1	7	914	Type 2	1	65	Type 3
R4	142	Type 1	9	851	Type 2	1	65	Type 3
R5	83	Type 1	6	1067	Type 2	4	72	Type 3

The results in Table 4.5 indicate that the presence or absence of CC driven changes in any *one* system forcing is not capable of changing the *Type* of any of the 3 case study STIs.

For Negombo lagoon, the results indicate that CC driven changes in wave conditions (in this case with an associated increase in *M)* have the largest impact on inlet stability, accounting for almost 70% of overall CC modified *r* value of 75 (by comparing *r* for R1, R3 and R4). Comparison of results for R1, R2 and R5 indicate that CC driven variations in riverflow and SLR both appear to have smaller but similar contributions to the overall CC effect on inlet stability (~10% contribution).

For Kalutara lagoon, the variations among *r* values computed for the 5 simulations are insignificant and stay within the $5 < r < 10$ range. Nevertheless, the variations in migration distance indicate that the phenomenon which contributes most to the 1210 m of migration due to combined CC forcing (R1) is CC driven variations in wave conditions (R3, 25% contribution to the overall migration distance).

At Maha Oya, while both the *r* value and time to closure for all simulations vary very little, the biggest drops in both diagnostics are attributable to CC driven variations in wave conditions (by comparing R1, R3 and R4).

The above results show that SLR appears not to be the main contributor to future CC driven variations in inlet stability and/or behaviour at any of the 3 case study STIs.

4.5   Conclusions

A snap-shot simulation approach using the process based coastal area morphodynamic model *Delft3D* has been applied to assess CC impacts on the stability of Small Tidal Inlets (STIs). The modelling approach was applied to three case study sites representing the main types of STIs: locationally and cross-sectionally stable inlets (Type 1, Negombo lagoon, Sri Lanka - permanently open, fixed in location); cross-sectionally stable, locationally unstable inlets (Type 2, Kalutara lagoon, Sri Lanka - permanently open, alongshore migrating); and locationally stable, cross-sectionally unstable inlets (Type 3, Maha Oya river, Sri Lanka - seasonally/intermittently open, fixed in location). Future CC modified wave and riverflow conditions were derived from a regional scale application of spectral wave models (WaveWatch III and SWAN) and catchment scale applications of a hydrologic model (CLSM) respectively, both of which were forced with IPCC

GCM output dynamically downscaled to ~65 km resolution over the study area with the Conformal Cubic Atmospheric Model CCAM.

The coastal impact model used in this study, *Delft3D*, was successfully validated for contemporary conditions using short-term hydrodyamic measurements and the general morphological behaviour observed in satellite images of the study sites. Subsequent CC impact simulations undertaken with the validated models forced with 2100 projected SLR, waves and riverflows indicate the following:

- None of the 3 case study STIs will change Type by the year 2100.

- By the end of the 21st century, the level of stability of the Negombo lagoon, as indicated by the Bruun criterion r, will decrease from 'Good' to 'Fair to poor'. The level of (locational) stability of the Kalutara lagoon, as indicated by the doubling of the annual migration distance, will also decrease. At Maha Oya river, the time till inlet closure will increase by more than 50%, indicating an increase in the level of stability of this inlet.

- CC driven variations in wave conditions, and resulting changes in the annual longshore sediment transport, is the main contributor to the overall CC effect on the stability of all 3 STIs. SLR and CC driven variations in riverflows play only a rather secondary role.

The results obtained by applying the 'data rich' approach to the 3 case study sites are, to a large extent, in agreement with those obtained from the more generically applied 'data poor' approach presented in Chapter 3.

References

Bamunawala, R. M. J., 2013. Impact of climate change on the wave climate of Sri Lanka. M.Sc Thesis, University of Moratuwa, Sri Lanka. 55p.

Boon, J. D., Byrne, R. J., 1981. On basin hypsometry and the morphodynamic response of coastal inlet sytems. Marine Geology, 40, 27-48.

Booij N., Ris, R.C., Holthuijsen L.H., 1999. A third generation wave model for coastal regions. Part 1: model description and validation. Journal of Geophysical Research 104 (C4), 7649-7666.

Charles, E., Idier, D., Delecluse, P., Deque, M., Le Cozannet, G., 2012. Climate change impact on waves in the Bay of Biscay. Ocean Dynamics, 62, 831-848.

Ducharne, A., Koster, R. D., Suarez, M. J., Stieglitz, M., Kumar, P., 2000. A catchment-based approach to modeling land surface processes in a GCM, Part 2, Parameter estimation and model demonstration. Journal of Geophysical Research, 105, 24823-24838.

Koster, R. D., Suarez, M. J., Ducharne, A., Stieglitz, M., Kumar, P., 2000. A catchment-based approach to modeling land surface processes in a GCM, Part 1, Model Structure. Journal of Geophysical Research, 105, 24809-24822.

Lesser, G., Roelvink, J.A., Van Kester, J.A.T.M., Stelling, G.S., 2004. Development and validation of a three-dimensional morphological model. Coastal Engineering 51, 883-915.

Mahanama, S. P. P., Koster, R. D., Reichle, R. H., Zubair, L., 2008. The Role of Soil Moisture Initialization in Sub-seasonal and Seasonal Streamflow Prediction - A Case Study in Sri Lanka. Advances in Water Resources, 31, 1333-1343.

Mahanama, S.P.P., Zubair, L., 2011. Production of streamflow estimates for the Climate Change Impacts on Seasonally and Intermittently Open Tidal Inlets (CC-SIOTI) Project. FECT Technical Report 2011-01: Foundation for Environment, Climate and Technology, Digana Village, October, 2011. 20p.

McGregor, J., Dix, M., 2008. An updated description of the conformal cubic atmospheric model. In: High resolution Simulation of the Atmosphere and Ocean (Eds. Hamilton, K., Ohfuchi, W.) Springer, pp. 51-76.

Ruessink, B.G., Ranasinghe, R., 2014. Beaches. In: Coastal environments and Global change (Eds. Masselink, G., Gehrels, R.). Wiley, pp. 149-176.

Tolman, H., 2009. User manual and system documentation of WAVEWATCH III™ version 3.14. NOAA/NWS/NCEP/MMAB Technical Note 276, 194 pp + Appendices. (URL http://polar.ncep.noaa.gov/waves/wavewatch/).

Wang, L., Ranasinghe, R., Maskey, S., van Gelder, P. H. A. J. M., Vrijling, J. K., 2015. Comparison of empirical statistical methods for downscaling daily climate projections from CMIP5 GCMs: a case study of the Huai River Basin, China. International Journal of Climatology, DOI 10.1002/joc.4334.

Wijeratne, E. M. S. (2002). Sea level measurements and coastal ocean modelling in Sri Lanka. Proceedings of the 1st scientific session of the National Aquatic Resources Research and Development Agency, Sri Lanka. 18p.

CHAPTER 5

A REDUCED COMPLEXITY MODEL TO OBTAIN RAPID PREDICTIONS OF CLIMATE CHANGE IMPACTS ON THE STABILITY OF SMALL TIDAL INLETS

5.1   Introduction

One of the main shortcomings of the process based snap-shot modelling approach discussed thus far in this thesis is its inability to provide any insights on the temporal evolution of STI stability in response to climate change (CC) modified system forcing. Another important issue is that the application of sophisticated process based models such as *Delft3D* under any situation requires a high level of specific expertise in using the model. Such expertise is uncommon among frontline coastal zone managers and planners who have to deal with CC impacts on coasts on a regular basis. In fact, frontline managers/planners appear to prefer simple models that can be used with a very limited knowledge of mathematical modelling but are still able provide reliable and rapid assessments of system behaviour (Ranasinghe at al., 2012; Wainwright et al., 2015). An easy-to-use reduced complexity (or aggregated) model that captures the physics governing inlet stability and is capable of providing rapid and reliable qualitative assessments of how inlet stability may evolve in time would address both of the above issues. Reduced complexity models have previously been successfully developed and used to simulate river morphodynamics (Murray and Paola, 1994, 1997), coastline change (Ashton and Murray, 2001; Ranasinghe et al., 2013) and inlet-basin systems (Stive and Wang, 2003; van Goor et al., 2003). In this chapter, a reduced complexity model to predict CC effects on STI stability is developed and applied to the three case studies STIs described in Chapters 3 and 4.

5.2   Governing processes

Under contemporary conditions, inlet stability is governed primarily by two processes: (1) annual ambient longshore sediment transport in the vicinity of the inlet (M), and (2) flow through the inlet during ebb. The latter will not only include the ebb tidal prism (volume of water flowing out of the estuary/lagoon due to tidal forcing alone) but also riverflow effects. For convenience, hereon, the ebb flow due to the combined effect of both of these processes will be referred to as tidal prism (P). CC driven variations in wave conditions and riverflow can modify both M and P in future.

5.2.1   Climate change and longshore sediment transport

At the most basic level, longshore sediment transport (LST) rate is a function of sediment size (D_{50}) and breaking wave height (H_b) and angle (θ_b). There are several bulk equations (e.g. CERC, 1984; Bayram et al., 2007) that express LST as a predominant function of these parameters. Kamphuis (1991) presented an LST equation which, apart from these parameters, also includes the effect of the surf zone bed slope (Eq. 5.1).

$$I_m = 2.27 H_b^2 T_p^{1.5} m_b^{0.75} D_{50}^{-0.25} sin^{0.6}(2\theta_b) \tag{5.1}$$

where: I_m is immersed mass of sediment transported alongshore (kg/s), H_b is the significant wave height at the breaker (m), θ_b is the wave angle at the break point (degrees), T_p is the peak wave period (s), $m_b = h_{br}/\lambda_{br}$ is the beach slope, h_{br} is the water depth at break point (m), λ_{br} is the distance from the shoreline to the break point (m), and D_{50} is the median grain size (m),

I_m is related to the volume via $Q_l = \frac{I_m}{(\rho_s - \rho)(1-p)}$, where: Q_l is the sediment transport volume (m³/s), ρ_s is the density of sand (kg/m³), ρ is the fluid density (kg/m³), p is porosity of sand.

This equation has been shown to perform remarkably well on large data sets (Kamphuis, 1991; Schoonees and Theron, 1996). Milhomens et al. (2013) re-assessed the commonly used CERC, Bayram and Kamphuis LST equations using a comprehensive database containing laboratory and field data and proposed new improved versions for all 3 equations. Among these however, the improved Kamphuis equation (Eq. 5.2) gave the best agreement with data.

$$I_{m,new} = 0.149 H_b^{2.75} T_p^{0.89} m_b^{0.86} D_{50}^{-0.69} sin^{0.5}(2\theta_b) \tag{5.2}$$

It is now well known that CC will result in sea level rise (SLR) and modified deepwater wave heights and directions around the world (Hemer et al., 2013). Such CC driven modifications in offshore wave conditions will also affect breaking wave properties H_b and θ_b. Furthermore, SLR will result in waves breaking closer to the shoreline and hence a steeper surf zone slope (higher m_b). All of these CC effects can be accounted for by using Eq. 5.2 with appropriate CC forcing.

5.2.2   Climate change and tidal prism

As mentioned above, the total ebb tidal prism can be divided into two parts:

$$P = P_t + R \tag{5.3}$$

where: P_t is the ebb flow due to tide only and R is the riverflow volume during ebb. In small estuary/lagoon system where it can safely be assumed that there is no phase lag in tidal elevation within the system (Keulegan, 1951), P_t can be expressed as:

$$P_t = A_b \times (2 \times a_b) \tag{5.4}$$

where: A_b is the surface area of the estuary/lagoon and a_b is the mean tidal amplitude inside the estuary/lagoon.

As STIs do not usually contain extensive tidal flats or salt marshes, it can be assumed that SLR will not change A_b. The lagoon tidal range a_b then is a function of the ocean tidal range and tidal attenuation across the inlet channel (Keulegan, 1951; van der Kreeke, 1988; Ranasinghe and Pattiaratchi, 2000). CC is not expected to affect ocean tides in any significant way and hence CC will only affect a_b via changes in tidal attenuation. Inlet channel dimensions such as length, width and depth will directly affect the degree of tidal attenuation (Keulegan, 1951).With CC, the competing effects of SLR and basin infilling (Ranasinghe et al., 2013) will result in increasing inlet depth by half the amount of SLR. This will decrease tidal attenuation, thus increasing P_t.

Keulegan (1951) presented an analytical solution to calculate tidal attenuation at inlets which can be used to calculate the CC modified P_t at STIs. The starting point for this approach is Eq. 5.5 which describes the difference in water level across a tidal inlet:

$$\frac{dh_b}{dt} = K\sqrt{h_o - h_b} \tag{5.5}$$

where: h_b is the water level inside the basin (m), h_o is the water level at the sea (m), K is the coefficient of filling or repletion, given by the Eq. 5.6 below:

$$K = \frac{T_t}{2\pi a_o} \frac{A}{A_b} \sqrt{\frac{2gra_o}{fL+mr}} \qquad (5.6)$$

where: T_t is the tidal period (s), a_o is the tidal amplitude at the sea (m), A is the cross-sectional area of the connecting channel (m^2), A_b is the basin surface area (m^2), g is the gravitational acceleration (m^2/s), r is the hydraulic radius of the channel (m), f is the friction coefficient, L is the length of the connecting channel (m), m is the coefficient resulting from the velocity distribution over the channel cross section.

As Eq. 5.5 is an implicit equation, after some mathematical manipulations and approximations for sinusoidal tidal oscillations, Keulegan (1951) arrived at Eq. 5.7, which is explicit.

$$\frac{a_b}{a_o} = sin\tau \qquad (5.7)$$

where: a_b is the tidal amplitude inside the basin (m), and $sin\tau$ is a function of K.

Keulegan (1951) further provided look-up tables for $sin\tau$ vs K. As the latter can easily be estimated for a given system from Eq. 5.6, a_b can then be calculated for a given ocean tidal amplitude a_o, which is generally known. This a_b can then be used, along with A_b, in Eq. 5.4 to compute P_t, which can be combined with a known riverflow (per ebb phase) to calculate P using Eq. 5.3. It is noted that Keulegan's (1951) approach does not allow a_b to be greater than a_o (i.e. $sin\tau$ always < 1) and therefore is not suitable for situations where tidal resonance may amplify a_b. However, for tidal resonance to occur the ratio length of basin/tidal wave length should be 0.25 (Dronkers, 1964). For STIs therefore this situation is unlikely to occur. For example, the length of basin/tidal wave length ratio of the 3 STIs investigated here ranges between 0.002 and 0.078.

5.3   The model

The basic structure of the reduced complexity model developed in this study is shown in Figure 5.1.

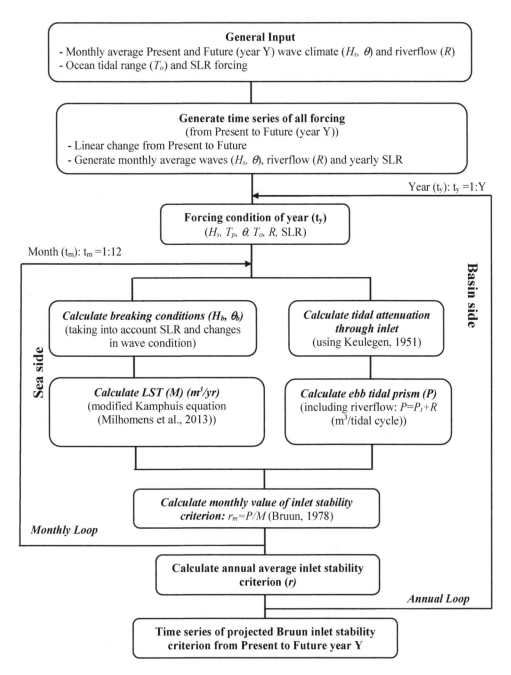

Figure 5.1. Basic model structure.

The starting point of the model is the generation of required model forcing time series. For this type of reduced complexity long term modelling approach, it is sufficient to use monthly averaged forcing (unless such averaging results in large under-predictions of especially peak values, in which case a shorter averaging window may be required). For simplicity, it can be assumed that all relevant contemporary system forcing that will be modified by CC (wave height, wave direction, riverflow and MWL) will linearly change from their present values to projected future values at a desired planning horizon (say, the year 2100). Sufficient historical data to generate the required monthly averaged time series of contemporary wave conditions and riverflow as well as downscaled projections of CC modified monthly averaged wave conditions and riverflow by 2100 will be required to generate the full forcing time series from the present to 2100 at monthly resolution. The SLR projections given by IPCC (2013) can be used directly to calculate an annually varying MWL time series for the simulation duration.

Breaking wave conditions, required by Eq. 5.2 to calculate LST, are rarely available from direct measurements or regional wave models. Therefore, offshore wave conditions (which are more readily available) at each monthly time step have to be first transformed to the nearshore until breaking occurs. In the model developed here, this is achieved by using Snell's law for wave refraction and the dispersion relation (Eq. 5.8) assuming that nearshore depth contours are parallel to the coastline and follow a Dean equilibrium profile corresponding to the measured D_{50} of the study area. This returns H_b, θ_b and the depth at which waves break (h_b). The breaker depth (h_b) is then used together with the concurrent MWL to calculate surf zone slope m_b required by Eq. 5.2. Subsequently, the monthly time series of H_b, θ_b and m_b are used in Eq. 5.2 to obtain monthly LST values for the entire simulation.

$$\omega^2 = gktanh(kh) \hspace{5cm} (5.8)$$

where: $\omega = \frac{2\pi}{T}$ is the angular frequency (rad/s), $k = \frac{2\pi}{\lambda}$ is the wave number (rad/m), λ is the wave length (m), and h is the water depth (m).

To calculate the monthly averaged time series of ebb tidal prism P, first the time series of P_t is required. In the model developed here, this is achieved by using Eq. 5.3-5.5 above. The time variation of P_t arises from the time variation of the repletion coefficient K (Eq. 5.6) which varies due to the SLR/basin infilling driven increase in inlet depth (relative to MWL). As SLR is only updated annually, P_t will only change every year. The riverflow R (per ebb phase) time series, which varies monthly, is then added to the P_t time series to produce monthly P values for the model duration.

Next, using the time series of P and LST values thus obtained, a monthly time series of the Bruun criterion r_m is generated using:

$$r_m = \frac{P}{M} \qquad\qquad (5.9)$$

Finally, the annual representative r value is obtained by taking the average of r_m over every year of the simulation. The model is very fast and produces a 100 year prediction of CC driven variation in r within a few seconds on a standard PC.

5.4   Model applications and Results

5.4.1   Model forcing

The above described model was applied from 2000 to 2100 to the 3 case study STIs described in Chapters 3 and 4: Negombo lagoon (Type 1); Kalutara lagoon (Type 2); and Maha Oya river (Type 3). Key system dimensions used are the same as those given in Table 3.1, while CC projections used for the year 2100 are those summarised in Section 4.3. A worst case SLR of 1m (by 2100 relative to the present) was applied at all 3 systems as in Chapter 4.The time series of system forcing for the present (taken as year 2000) and year 2100, the full 100 yr model forcing, and time series of P and M generated for the 3 case studies using the approaches described above in Section 5.3 are shown in Figures 5.2, 5.3 and 5.4 respectively.

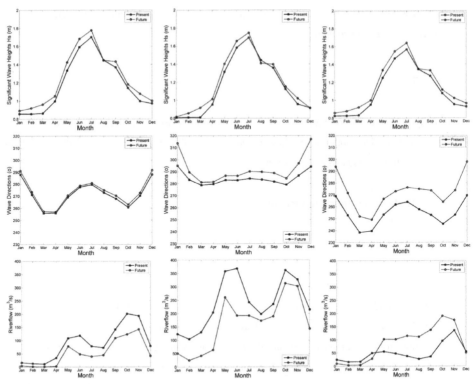

Figure 5.2. Contemporary and projected year 2100 forcing time series for Negombo lagoon (left), Kalutara lagoon (middle), and Maha Oya river (right); Significant wave height (H_s) at 10m depth (top), mean wave direction (θ) at 10m depth (middle), riverflow (bottom).

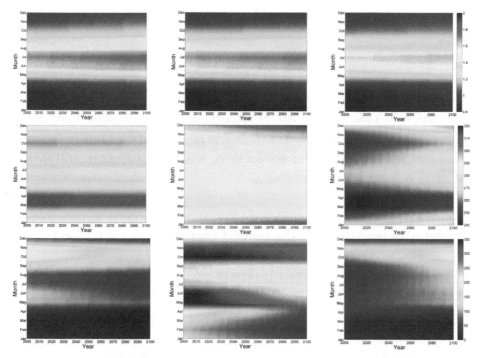

Figure 5.3. Full forcing time series generated for Negombo lagoon (left), Kalutara lagoon (middle), and Maha Oya river (right). Significant wave height (H) at 10m depth (top), mean wave direction (θ) at 10m depth (middle), Riverflow (bottom).

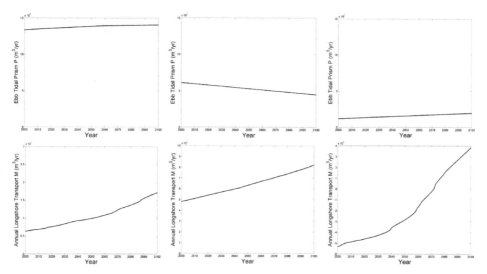

Figure 5.4. Time evolution of P (top) and M (bottom) for Negombo lagoon (left), Kalutara lagoon (middle), and Maha Oya river (right). M+ indicates southwards transport.

5.4.2   Model predictions of inlet stability

Model predicted temporal variation of inlet stability for the 3 systems are shown in Figure 5.5. The year 2100 r values and inlet Type predicted by the model are also compared with those predicted at the same time horizon by the process based snap-shot approach for data rich environments (Chapter 4) in Table 5.1. Note that, by necessity, the reduced complexity model completely depends on the relationship between r and inlet Type presented in Table 3.10 to predict future inlet condition, while the process based approach has the benefit of model predicted 2DH bed levels in addition to r values to predict inlet condition.

Table 5.1. Comparison of year 2100 r values and inlet Type predicted by the reduced complexity model and the process based snap-shot approach for data rich environments (Chapter 4).

	r	Trend of r change from present	Bruun classification	Inlet Type
Negombo lagoon (Present condition: Type 1- Permanently open, locationally stable, r =210)				
Reduced complexity model	82	decrease	Stable (fair to poor)	Type 1
Process based snap-shot model	75	decrease	Stable (fair to poor)	Type 1
Kalutara lagoon (Present condition: Type 2- Permanently open, alongshore migrating, r =13)				
Reduced complexity model	6	decrease	Unstable	Type 2 or 3
Process based snap-shot model	6	decrease	Unstable	Type 2
MahaOya river (Present condition:Type 3- Seasonally/intermittently open,locationally stable, r=2)				
Reduced complexity model	5	increase	Unstable	Type 3
Process based snap-shot model	5	increase	Unstable	Type 3

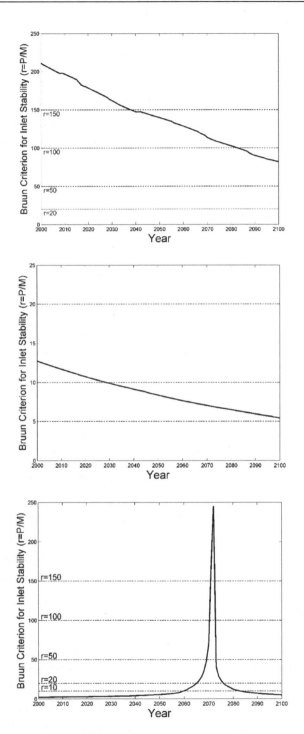

Figure 5.5. Model results for Negombo lagoon (top), Kalutara lagoon (middle), and Maha Oya river (bottom).

Negombo lagoon (Type 1)

Model predictions for 2100 match those of the corresponding process based snap-shot model predictions rather well in all diagnostics considered in Table 5.1. The model predicts that the level of stability (i.e. r value) of this system will gradually decrease in time (Figure 5.5, top). This is because while P remains practically unchanged from present to 2100, M increases by almost 200% (Figure 5.4, left). Table 5.1 indicates that Negombo lagoon will remain a locationally and cross sectionally stable Type 1 inlet by 2100, which is in agreement with the corresponding process based snap-shot model predictions presented in Chapter 4. Thus, it can be concluded with a high level of confidence that this system will remain as a Type 1 inlet during the 21st century. However, the level of stability decrease shown in Figure 5.5 (top) indicates that the inlet will deteriorate from a highly stable one ($r > 150$ at present) to an inlet with fair to poor stability ($50 < r < 100$) according the Bruun criteria for inlet stability.

Kalutara lagoon (Type 2)

Model predictions for 2100 match those of the corresponding process based snap-shot model predictions in all diagnostics considered in Table 5.1. In this case, too the model predicts that the level of stability will decrease over time (Figure 5.5, middle). This is due to the (almost) doubling of M while P also decreases during the 100 year simulation (Figure 5.4, middle). While the computed reduction of r from 13 to 6 is quantitatively small, this moves the inlet stability into the uncertain $5 < r < 10$ range where the inlet could be Type 2 or Type 3 as indicated in Table 3.10. Therefore, the only conclusion that can be drawn from the reduced complexity model, which completely depends on the r vs inlet Type relationship suggested in Table 3.10, is that the inlet could be either a Type 2 or Type 3 inlet by the end of the 21st century. On the other hand, the corresponding process based snap-shot model result for the end of the 21st century, which also gives an r value of 6, provides a more definitive indication regarding the inlet Type by 2100, in that it remains as an alongshore migrating Type 2 inlet. This highlights that when $5 < r < 10$, the reduced complexity model should ideally be used in conjunction with the process based snap-shot modelling approach to obtain reliable results.

Maha Oya river (Type 3)

In this case, model predictions exactly match those of the corresponding process based snap-shot model predictions in all diagnostics considered in Table 5.1, proving a high level of confidence that by the year 2100, the system will be a Type 3 inlet (as it is at present). However, due to its ability to provide the full temporal evolution of r over the 100 year study period, the reduced complexity model provides very interesting insights that the process based snap-shot modelling approach does not provide. Figure 5.5 (bottom) shows that r, which remains below 10 till about 2060, rapidly increases to > 150 around 2070, and then drops rapidly back to below 10 soon after 2080. This is because, while P remains almost unchanged through the 100 year simulation, M goes through a zero crossing around 2070 and changes direction from a northward transport to a southward transport due to the projected CC induced clockwise rotation of wave direction in the study area, particularly during the SW monsoon when energetic waves are present (Figure 5.3, right; top and middle panels). Due to this phenomenon, when northward M decreases rapidly towards zero from about 2060, r increases rapidly, and then drops again when southward M starts to increase during the 2070-2080 decade. This implies that Maha Oya inlet may turn into a permanently open, locationally stable Type 1 inlet for a few years during the first half of the 2070-2080 decade.

One could however argue that this modelled peaking of r in the 2070-2080 decade is due to the use of net annual longshore sediment transport to calculate r in the model. Indeed if gross annual longshore sediment transports were used in this calculation, the predicted peak r value (in 2072) would drop significantly (to below 20). However, while the initial publication by Bruun and Gerritsen (1960) regarding the P/M ratio advocates the use of gross annual longshore transport as an approximation for the '*total sand volume that drifts into the inlet channel*', later publications by Bruun (1978, 1991) suggest that it might be reasonable to use '*predominant annual littoral drift*' or '*net annual littoral drift*' to quantify r. In line with this suggestion, more recent numerical modelling studies (Ranasinghe et al., 1999; Nahon et al., 2010) have used the net annual longshore sediment transport to quantify r. In any case, it is unlikely that all the sand that is being transported alongshore by longshore currents gets deposited in the inlet channel. Some proportion will bypass the inlet via ebb deltas or submerged nearshore sand bars, while another part of the sand that enters the inlet during flood will simply be advected through the inlet to the basin and be deposited in flood deltas, thereafter becoming inactive. Thus, using the gross annual longshore sediment transport to calculate r is likely to result in an unrealistically low r value. On the other hand, in situations where the gross annual longshore sediment transport comprises significant transport volumes from both sides of the inlet, thus resulting in a very low net annual transport, an r value based on the net annual lonsghore sediment transport is likely to be unrealistically high. In such cases, it is probably best to calculate r using an M value that lies in between the gross and net annual transport rates. How exactly to determine the appropriate M value to use in this type of situations warrants further research but lies outside the scope of the present study.

5.4.3   Relative contributions of CC driven variations in system forcing to inlet stability

Apart from the simulations described above with complete CC forcing (simulation S1), 6 additional simulations where CC forcing was sequentially removed (simulations S2-S7) were undertaken to investigate the relative contribution of the different CC forcings to changes in future inlet stability. The CC forcings implemented in each simulation are shown in Table 5.2. This set of simulations was undertaken for all 3 case study sites and the results are shown in Figure 5.6.

Table 5.2. Forcing conditions implemented in the different CC forcing simulations. Subscript f indicates future conditions.

	SLR	$H_{S,f}$	θ_f	R_f
S1	X	X	X	X
S2	X	X		X
S3	X		X	X
S4	X	X	X	
S5	X			X
S6	X			
S7		X	X	X

Negombo lagoon (Type 1)

Figure 5.6 (top) shows that under all CC forcings considered, r always stays above 50 for this system, providing further confidence that Negombo inlet will remain as a Type 1 inlet during the 21st century. Significant (> 50) increases of the 2100 r value (relative to S1) are shown in S2, S5 and S6, implying that CC driven variations in wave direction most influences the model predicted decrease in the level of stability of this system.

Kalutara lagoon (Type 2)

In this case, r always stays in the 5-10 range (Figure 5.6, middle), suggesting (in conjunction with the corresponding process based snap-shot simulation) that this inlet will remain a Type 2 inlet in the 21st century. The largest change in the 2100 r value (relative to S1) is shown in S6, which is nevertheless a very small increase of 3, implying that the combined effect of CC driven variations in wave height, wave direction and riverflow plays a bigger role than SLR in determining the level of stability of the Kalutara lagoon inlet.

Maha Oya river (Type 3)

Figure 5.6 (bottom) shows that the rapid and short-lived system change to a Type 1 inlet predicted in S1 is absent in S2, S5 and S6. This shows the major influence that CC driven variations in wave direction have on determining the future behaviour of Maha Oya inlet. In fact, these results suggest that, if the CC driven variation in wave direction were to be absent, the behaviour of the Maha Oya inlet will remain more or less unchanged during the 21st century with a consistent r value of about 2. The peak r values reached around 2070 are much higher in S3 and S7 compared to all other simulations indicating that CC driven variations in wave height and SLR play a significant role in decreasing the level of stability of the inlet during the predicted peak stability period in the 2070-2080 decade.

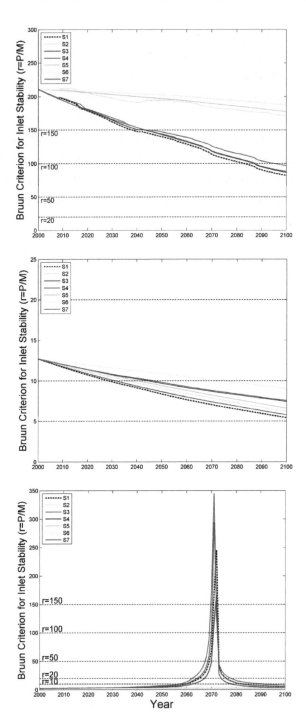

Figure 5.6. Relative contributions of CC forcing for the 3 systems as indicated in Table 5.2; Negombo lagoon (top), Kalutara lagoon (middle), Maha Oya river (bottom).

5.5   Conclusions

A reduced complexity model has been developed to obtain rapid assessments of the temporal evolution of the stability of small tidal inlets. The model is very easy to use and provides a 100 year projection of the Bruun inlet stability criterion r in less than 3 seconds on a standard PC.

Model applications to Negombo lagoon (Type 1 STI), Kalutara lagoon (Type 2 STI), and Maha Oya river (Type 3 STI) provided year 2100 predictions of inlet stability that are almost identical to those obtained with the process based snap-shot modelling approached described in Chapter 4. Model results show that:

- The Negombo lagoon inlet will remain as a Type 1 inlet during the 21st century. The level of inlet stability will deteriorate from a highly stable one ($r > 150$ at present) to an inlet with fair to poor stability ($50 < r < 100$). This decrease in inlet stability is largely due to CC driven variations in wave direction.

- The r value for Kalutara lagoon inlet will decrease into the uncertain range between 5 and 10 around 2030 and remain in that range until the end of the century. While the r vs inlet Type table presented in Chapter 3 would indicate that in this range of r the inlet could be either a Type 2 or Type 3 inlet, the corresponding process based snap-shot model results show that Kalutara lagoon will be a Type 2 inlet by the end of the 21st century. The combined effect of CC driven variations in wave height, wave direction and riverflow plays a bigger role than SLR in determining the level of stability of the Kalutara lagoon inlet. In general, these results indicate that for alongshore migrating Type 2 inlets, the reduced complexity model should ideally be used in conjunction with the process based snap-shot modelling approach to obtain reliable results if at any stage $5 < r < 10$.

- By the year 2100, Maha Oya inlet will be a Type 3 inlet (as it is at present). In contrast to the process based snap-shot modelling approach which only predicts the inlet condition at 2100, the reduced complexity model, due to its ability to provide the full temporal evolution of r over the 100 year study period, indicates that this system may turn into a permanently open, locationally stable Type 1 inlet for a few years during the 2070-2080 decade. This change is directly related to the CC driven variation in wave direction. CC driven variations in wave height and SLR play a significant role in decreasing the level of stability of the inlet during this peak inlet stability period in the 2070-2080 decade.

In general, model results show that CC is unlikely to change STI Type, unless dominant wave direction changes from one side of the shore normal to the other. If this occurs either at Type 2 or 3 STIs which have poor stability ($r < 20$), the system could temporarily turn into a Type 1 STI due to the associated reduction in annual longshore sediment transport. Model results also show that the CC effect that will affect STI stability most is increases/decreases in annual longhsore sediment transport due to CC driven variations in wave direction.

References

Ashton, A. D., Murray, A. B., Arnoult, O., 2001. Formation of coastline features by large-scale instabilities induced by high-angle waves. Nature, 414, 296-300.

Ashton, A. D., Murray, A. B., 2006. High angle wave instability and emergent shoreline shapes: 1. Modeling of sand waves, flying spits, and capes. Journal of Geophysical Research, 111, F04011, doi: 10.1029/2005JF000422.

Bayram, A., Larson, M., Hanson, H., 2007. A new formula for the total longshore sediment transport rate. Coastal Engineering, 54(9), 700-710, 2007.

Bruun, P., 1978. Stability of tidal inlets - theory and engineering. Developments in Geotechnical Engineering. Elsevier Scientific, Amsterdam, 510p.

CERC, 1994. Shore Protection Manual, volume I and II. Coastal Engineering Research Center, USACE, Vicksburg, 1984.

Dronkers, J. J., 1964. Tidal Computations in Rivers and Coastal Waters. North-Holland, Amsterdam; Interscience (Wiley), New York, 1964, 518 p.

Kamphuis, J. W., 1991. Alongshore sediment transport rate. Journal of Waterway, Port, Coastal and Ocean Engineering, 117(6), 624, 1991.

Keulegan, G. H., 1951. Third Progress Report on Tidal Flow in Entrances, Water Level Fluctuations of Basins in Communication with Seas, Report No. 1146, National Bureau of Standards, Washington, DC.

Milhomens, J., Ranasinghe, R., van Thiel De Vries, J., Stive. M., 2013. Re-evaluation and improvement of three commonly used bulk longshore sediment transport formulas. Coastal Engineering, 75, 29-39.

Murray, A. B., Paola, C., 1994. A cellular model of braided river. Nature, 371, 54-57.

Murray, A. B., Paola, C., 1997. Properties of a cellular braided-stream model. Earth Surface Processes and Landforms, 22, 1001-1025.

Ranasinghe, R., Pattiaratchi, C., 2000. Tidal Velocity Asymmetry at Inlets located in Diurnal Tidal Regimes. Continental Shelf Research, 20, 2347-2366.

Ranasinghe, R., Callaghan, D., Stive, M., 2012. Estimating coastal recession due to sea level rise: Beyond the Bruun Rule. Climatic Change 110, 561-574.

Ranasinghe, R., Duong, T.M., Uhlenbrook, S., Roelvink, D., Stive, M., 2013. Climate change impact assessment for inlet-interrupted coastlines. Nature Climate Change, 3, 83-87, DOI.10.1038/NCLIMATE1664.

Schoonees, J.S., Theron, A.K., 1993. Review of the field-data base for longshore sediment transport. Coastal Engineering, 19 (1-2), 1-25.

Stive, M.J.F., Wang, Z.B., 2003. Morphodynamic modelling of tidal basins and coastal inlets. Advances in Coastal Modelling (Ed. C. Lakhan). Elsevier Science B.V, pp. 367-392.

Van Goor, M.A., Zitman, T.J., Wang, Z.B., Stive, M.J.F., 2003. Impact of sea-level rise on the morphological equilibrium state, Marine Geology, 202, 211-227.

van de Kreeke, J., 1988. Hydrodynamics of tidal inlets. In: Hydro-dynamics and Sediment Dynamics of Tidal Inlets (Eds. Aubrey, D.G., Weishar, L.). Springer, Berlin, pp. 1-23.

Wainwright, D. J., Ranasinghe, R., Callaghan, D. P., Woodroffe, C. D., Jongejan, R., Dougherty, A, J., Rogers, K., Cowell, P. J.., 2015. Moving from Deterministic towards Probabilistic Coastal Hazard and Risk Assessment: Development of a Modelling Framework and Application to Narrabeen Beach, New South Wales, Australia. Coastal Engineering, 96, 92-99.

CHAPTER 6

GENERAL CONCLUSIONS

The general conclusions of this study are formulated below as responses to the 6 research questions posed in Chapter 1.

Research Question **1: Can a process based coastal area model be used to assess CC impacts on STIs?**

Yes. Although it is presently not feasible to apply a process based coastal area model to simulate the temporal morphological evolution of STIs over typical CC impact assessment time scales (e.g. 100 yrs), 1 year long snap-shot simulations that take into account the annual variability in wave conditions/riverflow can be used to assess CC impacts on STIs. Chapter 2 of this study has resulted in the development of 2 different process based snap-shot modelling approaches for data poor and data rich environments. The data poor approach uses schematized flat bathymetries that follow real world STIs and CC forcing derived from freely available coarse resolution global models while the data rich approach requires detailed bathymetries and downscaled CC forcing.

Research Question **2: Can an easy-to-use reduced complexity model be developed to obtain rapid assessments of the temporal evolution of STI stability under CC forcing?**

Yes. Using existing knowledge and physical fomulations, a reduced complexity model that can simulate the temporal evolution of STI stability based on the Bruun inlet stability criterion r (= P/M) value was developed and successfully demonstrated at 3 case study sites representing the 3 main STI types in Chapter 5 of this study. The model is capable of simulating 100 years in under 3 seconds on a standard PC.

Research Question 3: **Is there a link between STI Type and the Bruun inlet stability criterion r that could aid in classifying different STI responses to CC?**

Yes. Based on strategic process based snap-shot simulations undertaken in Chapter 3 of this study, a clear link between STI Type and r was established as shown in Table 6.1 below:

Table 6.1. Classification scheme for inlet Type and stability conditions.

Inlet Type	$r = P/M$	Bruun Classification
Type 1	> 150	Good
	100 - 150	Fair
	50 - 100	Fair to Poor
	20 - 50	Poor
Type 2	10 - 20	Unstable (open and migrating)
Type 2/3	5 - 10	Unstable (migrating or intermittently closing)
Type 3	0 - 5	Unstable (intermittently closing)

Research Question 4: **Will CC change STI Type?**

Mostly, No. The application of the data poor approach with the maximum projected ranges of CC driven variations in wave conditions and riverflows for the end of the 21st century, together with the worst case SLR of 1m, to schematized bathymetries representing the 3 main STI types indicates that, in general, Type 1 (Permanently open, locationally stable inlets) and Type 3 (Seasonally/ intermittently open, locationally stable inlets) STIs will not change Type by 2100. Application of the data rich approach and the reduced complexity model to the case study sites confirms this conclusion. The predictions made by the 3 different modelling approaches for Type 2 STIs are somewhat inconsistent in that while the data poor approach suggests a Type change to a Type 1 STI when CC results in a decreased annual longshore sediment transport rate, the case study applications of the other two approaches predict no Type change under any CC forcing scenario. This is, however, likely due to the difference between CC forcing adopted in the data poor approach (global maximum range of CC driven variations in forcing) and the other two approaches (site specific, downscaled CC forcings which are not as extreme as those used in the data poor approach).

***Research Question* 5: How will CC affect the level of stability of STIs?**

In general, SLR appears not to be the main driver of change in the level of STI stability. The stability of Type 1 STIs is governed by future changes in annual longshore sediment transport due to CC driven variations in wave directions. The level of stability of Type 2 and Type 3 STIs are equally affected by CC driven variations in wave conditions and riverflow. In these systems, concurrent increases (decreases) in longshore sediment transport and decreases (increases) in riverflow result in decreasing (increasing) the level of inlet stability.

***Research Question* 6: What basic guidelines can be given to coastal zone managers on how to assess CC impacts on STIs to inform CC adaptation strategies?**

For Type 1 and Type 3 STIs, regardless of whether they are in data poor or data rich environments, it is recommended that the reduced complexity model presented in Chapter 5 be first applied. If model results suggest an inlet Type change at any time, then it is recommended that process based snap-shot modelling be undertaken following the data poor or data rich approach as feasible for the study area. If however, apart from knowledge of possible future STI Type changes, insights on time till inlet closure (for a Type 3 STI) are also required, then it is essential that the (data poor or data rich) process based snap-shot modelling approach be adopted.

For Type 2 STIs also, regardless of whether they are in data poor or data rich environments, an initial application of the reduced complexity model presented in Chapter 5 is recommended. If the model result indicates that r always remains between 10 and 20, then application of the process based snap-shot modelling approach is only required if insights on future migration distance are required. However, if the reduced complexity model result indicates that r could at any time drop below 10, then it is essential that the (data poor or data rich, depending on which is feasible in the study area) process based snap-shot modelling approach be adopted.

Acknowledgements

The last few years in Delft have been a very rewarding and interesting journey for me. I am very grateful to many people for their friendship, support and contribution during this journey. First of all, I would like to thank Professor Dano Roelvink and Professor Rosh Ranasinghe for giving me the great opportunity to undertake very interesting research under their supervision. First being your MSc student and then your PhD student, I had the great fortune to work closely with you and learn so many new things.

Dano, I still remember after my MSc, you asked me "what is next?" and suggested that maybe I can do something later on tidal inlets in VietNam. At that time, I did not think it would happen but in the end that is indeed what happened - not on inlets in VietNam but on inlets in Sri Lanka. Thank you for your guidance and always spending your valuable time on my project whenever I needed your help. With your great knowledge and experience in process based modelling, you have helped me quickly solve several critical modelling problems that I would have otherwise had to struggle with for a very long time. You are a magician with numerical models! Thank you also for teaching me very early-on how to prepare for meetings, which has helped me ever since in conveying key messages efficiently at meetings. I owe you a very special thank you for always having the patience to answer my 'last question' with a smile!

Rosh, thank you for your confidence in me from the beginning to the end. I consider myself very fortunate to have you as a supervisor. You taught me how to simplify complex things in a very logical and detailed way. Thank you for the inspiration, unlimited encouragement and support, patient guidance and for always being available and giving me your time. I also greatly appreciate you for trusting me enough to let me freely pursue my own ideas and do things in my often stubborn way, even when my rate of progress was not upto your expectations. I must say at times I found it frustrating to answer (or find answers) to your very specific questions on details, but I now realise how valuable your attention to detail is. It has greatly helped me to critically evaluate my own work. So thank you for sometimes giving me a hard time and for your final push to wrap-up everything during the last few months of my PhD. But sorry, I cannot agree with you that "Billie Jean" is the best song ever!

I thank all of the committee members for examining my Thesis and the inspiring discussions and valuable comments which made me see some things from a different angle which only added to my appreciation of the topic of my thesis.

To UNESCO-IHE staff, colleagues and friends, thank you for all the support and life-sharing moments, thanks for creating the always very nice, warm and friendly atmosphere in our institute which feels so welcoming and like a home away from home. Special thanks to the CSEPD chair group for exchanging coastal knowledge and for all the group dinners, proposition meetings, colloquia etc which were a lot of fun. A special thank you to the Head of Water Science and Engineering Department, Professor Arthur Mynett. Thank you very much for your kindness, caring and encouragement, and sometimes going out of your way to solve students' problems. I really enjoyed your musical performances, along with Dano and other IHE staff, during our many UNESCO-IHE parties, and I hope you will continue to play lead guitar for many many more years and keep singing "Money for Nothing" exactly like Mark Knopfler! Jaap Kleijn, and Gordon de Wit, thank you for always helping me out with any IT problems I had at UNESCO-IHE. Tonneke Morgenstond, thank you very much for always being friendly and quickly attending to any administrative issues I came to you with.

I would also like to thank Deltares for hosting me during my entire PhD. Without access to the Deltares Cluster, my PhD, with its many Delft3D morphodynamic simulations, simply would not have been possible. A big thank you for Deltares staff, colleagues and friends for all the support through the years and providing a very nice working atmosphere. Thank you Professor Zheng Wang, Dirk Jan Walstra, and Arjen Luijendijk for your help, guidance and valuable discussions that helped me to get past some critical barriers. Thanks also to Herman Kernkamp, Erik de Goede, Rob Uittenbogaard, Mohamed Nabi, Robert Leander, Adri Mourits, Thieu van Mierlo, Stef Hummel, Jan van Kester, Bert Jagers, Agnes Arkesteijn, Radha Ramoutar and Patricia Tetteroo for your help, hospitality and advice. Rob and Herman, thank you for the lifts to town when it was raining, and Nabi thank you so much for staying late sometimes so that I could work late during the last few months of my PhD.

My dear friends Hien, Kitty, Thuy, Shaimaa, Alida and Janaka - thanks for all your help, support and caring in many different ways and for being there whenever I needed help through the years.

This PhD study was part of the UPARF project CC-SIOTI, which involved many research groups from around the world: University of Moratuwa, University of Peradeniya, and the Foundation for Environment Climate and Technology in Sri Lanka; Asian Institute of Technology, Thailand; CSIRO, Australia; Lund University, Sweden; and Deltares and UNESCO-IHE in The Netherlands. I would like to thank all project partners for promptly providing me all data/information I required for my study. I also thank you for your hospitality and generosity whenever I travelled to your countries for field work/meetings/symposia. It was great to get to know so many different people and cultures (and cuisines!) within my PhD project.

Finally my greatest gratitude to my family, Jniu, Amz-Gau, ChuRongBe, BGModen and Bebe. Thank you for always being there for me ("I know you will catch me if I fall") with your endless, and unconditional love and support. Without you I would not have been able to travel this far.

Trang Minh Duong

November 2015

About the author

Trang Minh Duong was born in Hanoi, VietNam. She obtained her entire undergraduate education at the Water Resources University, Hanoi. Subsequently she completed her MSc at UNESCO-IHE, Delft, The Netherlands. During her MSc she specialised in the Coastal Science Engineering and Port Development program within the Water Science and Engineering Department at UNESCO-IHE. For the MSc research component, Trang undertook a Deltares funded study on *The Hydrodynamics of fringing reef systems* which involved one of the first applications of *Xbeach* to reef environments. Thereafter, she commenced her PhD research which constituted the central part of a UNESCO-IHE lead, multi-stakeholder project CC-SIOTI that involved several research groups from Sri Lanka (University of Moratuwa, University of Peradeniya, and the Foundation for Environment Climate and Technology), Thailand (Asian Institute of Technology), Australia (CSIRO), and The Netherlands (Deltares). During her entire PhD candidature Trang was hosted by Deltares. The project was supported via the UPARF research program funded by the Dutch Foreign Ministry (DGIS) and UNESCO-IHE under DUPC programmatic funding. She has published several journal and conference articles to date.

Journal Publications

R. Ranasinghe, **T. Duong**, S. Uhlenbrook, D. Roelvink and M. Stive. 2013. Climate Change impact assessment for inlet-interrupted coastlines. *Nature Climate Change,* Vol. 3, 83-87.

A. van Dongeren, R. Lowe, A. Pomeroy, **T. Duong**, D. Roelvink, G. Symonds and R. Ranasinghe. 2013. Numerical modelling of Low-Frequency wave dynamics over a fringing coral reef. *Coastal Engineering,* Vol. 73, 178-190.

T. Duong, R. Ranasinghe, A. Luijendijk and D. Roelvink. 2012. Climate Change impacts on the stability of small tidal inlets. *Journal of Ocean and Climate Systems,* Vol. 3,163-171.

Trinh, T., Nguyen, H., Dinh, Q., **Duong, T**. and Nguyen, K. 2006. A studied of mechanism of the rainfall induced slope failures. *Science and Technology Journal of Agriculture and Rural Development:* 50-55.

Conference Proceedings

T. Duong, R. Ranasinghe, P. Dissanayake, A. Luijendijk, D. J. R. Walstra and D. Roelvink. 2015. Climate change driven morphological behaviour at small tidal inlet systems. *Proceedings of the Coastal Sediments 2015, San Diego, USA.*

P. Dissanayake, **T. Duong**, H. Karunarathna and R. Ranasinghe. 2015. Sediment dynamics of Negombo Lagoon Outlet, Sri Lanka. *Proceedings of the Coastal Sediments 2015, San Diego, USA.*

T. Duong, R. Ranasinghe, A. Luijendijk, D.J.R. Walstra and D. Roelvink. 2014. A qualitative assessment of climate change impacts on the stability of small tidal inlets via schematized numerical modelling. *Proceedings of the ICCE conference 2014, South Korea.*

T. Duong, R. Ranasinghe, A. Luijendijk, D. J. R. Walstra and D. Roelvink. 2014. Climate change and morphologic responses of small tidal inlets. *Proceedings of the International Symposium on Impact of Climate Change on the Coastal zone, Colombo, Sri Lanka, pp. 24-30.*

R. Ranasinghe, R. Holman, M. A. de Schipper, T. Lippmann, J. Wehof, **T. Duong**, D. Roelvink and M. J. F. Stive. 2012. Quantifying morphological recovery time scales using Argus video imaging: Palm beach, Sydney and Duck, NC. *Proceedings of the 33rd International Conference on Coastal Engineering (ICCE) 2012, Santander, Spain.*

A. van Dongeren, R. Lowe, A. Pomeroy, **T. Duong**, D. Roelvink, R. Ranasinghe and G. Symonds. 2012. Modelling infragravity waves and currents across a fringing reef. *Proceedings of the 33rd International Conference on Coastal Engineering (ICCE) 2012, Santander, Spain.*

T. Duong, Van Dongeren, A., Lowe, R., Roelvink, D., Ranasinghe, R. and Symonds, G. 2012. Modelling infragravity waves across the Ningaloo (Western Australia) fringing reef. AGU Ocean Sciences meeting.

T. Duong, R. Ranasinghe, A. Luijendijk, A. Dastgheib and D. Roelvink. 2012. Climate change impacts on the stability of small tidal inlets: A numerical modelling study using the realistic analogue approach. *Proceedings of COPEDEC VIII, Chennai, India, pp. 594-602.*

T. Duong and R. Ranasinghe. 2011. Climate change impacts on the stability of small tidal inlets. *Proceedings of Symposium on Climate Change Impacts on Small Tidal Inlets 2011, Bangkok, Thailand, pp. 8-15.*

A. van Dongeren, R. Lowe, A. Pomeroy, **T. Duong**, D. Roelvink, R. Ranasinghe and G. Symonds. 2010. Modelling infragravity waves and currents across a fringing reef: Ningaloo Reef, Western australia. *AGU Fall Meeting, San Francisco, USA.*

Lowe, R., Symonds, G., Moore, C., Ivey, G., Pattiaratchi, C., Van Dongeren, A., **T. *Duong* and *Roelvink, D.* 2010. *Dynamics of infragravity waves in fringing reef systems. Proceedings of the Australian Wind Waves Research Science Symposium, Gold Coast, Queensland, Australia, pp. 15-18.*

T - #0434 - 101024 - C132 - 244/170/7 - PB - 9781138029446 - Gloss Lamination